JOHN PLAYFAIR

OUTLINE

OF

NATURAL PHILOSOPHY

BEING

HEADS OF LECTURES

DELIVERED IN THE UNIVERSITY OF EDINBURGH

Volume 2

Elibron Classics
www.elibron.com

Elibron Classics series.

ISBN 1-4021-6418-1 (paperback)
ISBN 1-4021-1421-4 (hardcover)

This Elibron Classics Replica Edition is an unabridged facsimile of the edition published in 1816 by Archibald Constable & Co., etc., Edinburgh.

OUTLINES

OF

NATURAL PHILOSOPHY.

OUTLINES

OF

NATURAL PHILOSOPHY,

BEING

HEADS OF LECTURES

DELIVERED IN THE UNIVERSITY OF EDINBURGH,

BY

JOHN PLAYFAIR,

PROFESSOR OF NATURAL PHILOSOPHY IN THE UNIVERSITY OF EDINBURGH,
FELLOW OF THE ROYAL SOCIETY OF LONDON,
SECRETARY OF THE ROYAL SOCIETY OF EDINBURGH,
AND MEMBER OF THE ROYAL MEDICAL AND PHYSICAL SOCIETIES
OF EDINBURGH.

VOL. II.

SECOND EDITION.

EDINBURGH:
Printed by Abernethy & Walker,
FOR ARCHIBALD CONSTABLE AND COMPANY, EDINBURGH,
AND LONGMAN, HURST, REES, ORME & BROWN,
AND CADELL & DAVIES, LONDON.

1816.

CONTENTS

OF

VOLUME SECOND.

PART I.

ASTRONOMY.

SECT. 8.

OUTLINES

OUTLINES

OF

NATURAL PHILOSOPHY.

ASTRONOMY.

PART I.

PHENOMENA OF THE HEAVENS.

SECTION I.

OF THE FIXED STARS, AND THE CIRCLES OF THE SPHERE.

1. WHEN we look at the heavens in a clear night, all the stars appear at the same distance from us, and therefore seem to be situated in the surface of a sphere having the eye for its centre.

2. If we take a view of the stars again, in the course of the same night, we find that they retain their positions relatively to one another, but have changed them relatively to the earth,

having all made a progress toward the west, where some have set, while others have risen in the east. More attentive and continued observation shews, that they all describe circles having the same point for their Pole, or that from which the parts of the same circumference are all at an equal distance.

The stars, therefore, move exactly as if they were fixed in the surface of a sphere, which revolved on its axis in a space of time nearly equal to 24 hours. This motion, which is common to all the heavenly bodies, is called the *Diurnal Motion.*

3. If we observe the place of the Moon on two successive nights, we shall find that she changes her position among the stars, and advances to the eastward at the rate nearly of 13 degrees in 24 hours; so that she completes the circle of the heavens in 27 days nearly.

a. Not only the Moon, but the Sun, and ten other stars, (five of them visible only with the telescope), besides the diurnal motion, have motions eastward, relatively to the other stars.

b. The stars, which have no sensible motion relatively to one another, are called *Fixed Stars;* the others are called *Planets,* and will afterwards be more fully considered. Besides the planets, there are also stars which do not remain permanently visible, and which have motions of their own. These are the

the *Comets*; and are also to be considered hereafter.

4. The Earth is a body bounded by a surface, nearly spherical, the diameter of which is exceedingly small, compared with the distance of the fixed stars.

a. The round or globular figure of the earth is inferred from many phenomena.

1. From the appearance of ships at different distances from one another, and of the land when approached from the sea.

2. From the circum-navigation of the globe.

3. From the conical shadow of the earth, as seen in eclipses of the moon.

b. The circumference of the earth is nearly 25020 miles; its diameter 7910 miles, and a degree something more than 69.

These measures are not given as correct. The method of ascertaining them with precision will be afterwards explained.

c. The smallness of the diameter of the earth compared with the distance of the fixed stars, is inferred from this, that the circles which those stars appear to describe, are the same to spectators in all parts of the earth, and also that the angles subtended by the distance of any two stars from one another, is the same at all points on the earth's surface.

Since

Since the diameter of the earth is very small in respect of the distance of the stars, we may suppose the centre of the sphere in which these stars appear to be placed, to be the centre of the earth, and not the eye of the observer.

5. The straight line which passes through the pole of the Northern Hemisphere, and through the centre of the earth, is called the *Axis of the Heavens*, and is the line about which the heavens appear to revolve. The point in which this line seems to meet the superficies of the Southern Hemisphere, is called the South Pole of the Heavens, as the other is called the North Pole.

The Poles are also distinguished by the names of *Arctic* and *Antarctic*.

6. As it is necessary, in order to ascertain the position of any point on the surface of a sphere, to refer it to two planes given in position ; so, in order to determine the places of the stars for any moment, and thereby to ascertain the laws of their motion, it is necessary to suppose certain planes to pass through the centre of the celestial sphere, to which these places may be referred.

This gives rise to what is called the Doctrine of the Sphere, and the application of spherical trigonometry to astronomy.

7. A

7. A circle, the plane of which passes through the centre of the earth, and is perpendicular to the axis above defined, is called the *Equator*. The line in which this plane cuts the surface of the earth is called the *Equinoctial*.

a. The circles described by the stars in their diurnal motion, are all parallel to the equator.

8. If any plane pass through the poles of the celestial sphere, the circle in which it cuts the surface of that sphere is called a *Meridian Circle*; and the section of the same plane with the surface of the earth, is called the *Meridian* of any of the points through which it passes.

a. The planes of all the Meridians cut the plane of the Equator at right angles.

9. If at any place on the earth's surface, the *direction of gravity*, (that is, the direction in which bodies fall to the ground, or the line in which the plummet hangs), be produced upward, the point in which it cuts the celestial sphere is called the *Zenith* of the place; and if the same line be produced downward, the point in which it cuts the celestial sphere is called the *Nadir*. If a plane perpendicular to this line pass through the place itself, the circle in which it cuts the heavens is called the *Horizon*.

a. The

a. The direction of gravity passes nearly through the centre of the earth.

b. The Zenith and the Nadir are the Poles of the horizon.

c. If a plane parallel to the plane of the horizon of any place, pass through the centre of the earth, it will cut the celestial sphere in a circle hardly distinguishable from the former. It is called the *Rational*, and the other the *Sensible* Horizon.

d. The Horizon of any place divides the visible from the invisible Hemisphere. This supposes the eye to be situated on the surface of the earth; if it is elevated above the surface, the apparent horizon is a circle parallel to the former, but lower, being the base of a cone which has the eye of the spectator for its vertex, and of which the superficies touches the superficies of the earth all round.

10. Circles, the planes of which pass through the Zenith and Nadir of any place, are called *Vertical Circles*, and are perpendicular to the horizon of the place.

a. The Meridian of any place, is one of the vertical circles of that place.

b. The vertical circle that is at right angles to the plane of the meridian, is called the *Prime Vertical.*

a. If

c. If a vertical circle pass through a star, (or any point in the heavens), the arch of that circle intercepted between the star and the horizon is called the *Altitude* of the star; and the arch of the horizon intercepted between the said vertical and the meridian is called the *Azimuth* of the star.

d. Vertical Circles are also called *Circles of Azimuth.*

e. The altitude of a star above the horizon, is greatest when it is on the meridian; and a star has equal altitudes when it is at equal distances from the meridian on either side, or when its eastern and western azimuths are equal.

11. All the circles hitherto defined, are determined in their position, either by the diurnal motion of the heavens, or by the direction of gravity.

12. If a meridian circle pass through any star, the arch of that circle intercepted between the star and the Equator is called the *Declination* of the star; and if it pass through the zenith of any place, the arch intercepted between that zenith and the equator is called the *Latitude* of the place.

13. The Latitude of any place is equal to the altitude

altitude or elevation of the pole above the horizon of that place.

14. The elevation of the pole at any place is found, by observing one of the stars, which are so near it as not to set, when it passes the meridian above, and again when it passes the meridian under the pole; that is, by observing its greatest and its least altitudes above the horizon; half the sum of the two altitudes of the star, is the elevation of the pole, or the Latitude of the place.

For the star is always at the same distance from the pole; so that being on the meridian, it was as much above the pole in the first observation, as it was below in the second.

a. Those stars never set at any place of which the polar distance is equal to the latitude of the place.

b. This observation requires the use of an instrument which can be placed accurately in the plane of the meridian, such as the Astronomical Circle, the Quadrant, &c. or one like HADLEY'S Sextant, which can determine the greatest or least altitude of a star above the horizon, without any previous inquiry into the position of the meridian.

15. To

15. To find the Meridian Line, or that in which the plane of the meridian intersects the horizon of any place.

Observe the altitude of a star when it is on the east side of the meridian, and mark on the horizon the point which is in the same vertical circle, or which has the same azimuth with it. Observe the star when it has the same altitude again on the west side of the meridian, and mark in like manner the point on the horizon which is in the same vertical circle with it. The line that bisects the angle made by lines drawn from the place of observation to the two points thus marked on the horizon, is the Meridian Line.

For when the altitudes of a star on opposite sides of the meridian are equal, its azimuths, or the angles which its verticals make with the meridian, are equal also.

16. The meridian and the latitude being thus found, if the meridian altitude of any star be observed, its distance from the pole, or from the equator, is determined.

a. For the altitude of the equator, or of the point where the equator cuts the meridian, is known, being the same with the complement of the latitude; and the difference between the meridian altitude of the star and of the equator is the declination of the star.

b. If

b. If the meridian altitude of the star is less than that of the equator, its declination is on the opposite side of the equator from the zenith of the place, otherwise on the same.

Thus the places of the stars, as north or south of a given plane, that of the equator, are determined; but those places, to be fully known, must be determined in the direction of east and west, in respect of a given circle at right angles to the equator. Now, of the circles at right angles to the equator, or the circles of declination, no one is fixed in its position, but all of them revolve uniformly with the heavens; so that it is only by the computation of time that one of them can be distinguished from another.

The circle of declination which passes through the point in the equator which the sun occupies at the vernal equinox, has some advantages above the rest, as a line to which the stars are to be referred. It is called the *Equinoctial Colure*, and its position, with respect to the meridian of a given place, for a given instant, may be determined by means of a clock, regulated according to the following method.

17. The time from a star being on the meridian, to its next coming to the meridian, is always of the same length; it is called a Siderial Day.

The

The better the machines for measuring time are constructed, the more nearly are they found to agree with the motion of the heavens. By this the uniformity of both motions is ascertained.

The siderial day is divided into 24 hours; and as the whole circumference of the equator passes in that time over the meridian of any place, an arch of 15° passes in an hour; of 1°, in four minutes; of 1′ in 4 seconds of time; and so in proportion.

18. To compare the period of a clock with the length of the siderial day.

Observe the hour at which a star comes to the meridian on any two successive days; the interval ought to be 24 hours; if it is more, the clock goes too fast; if less, too slow. By repeating these observations day after day, the rate of the clock, or the quantity by which it daily advances on true time, or falls short of it, will be found. By shortening the pendulum if the clock goes too slow, or lengthening it if the clock goes too fast, the motion of the clock may be brought nearly to agree with that of the stars.

19. The period of the clock being thus adjusted, it is required to find what hour the clock marks, when the point of the vernal equinox is on the meridian of a given place.

Observe

Observe the hour by the clock when the sun's centre is on the meridian; and observe also the declination of the sun at the same instant. Let the angle at which the sun's path cuts the equator, or what is called the Obliquity of the Ecliptic, $=23^\circ.27'.30''$;

find an arch ϕ, such that $\sin \phi = \frac{\tan \text{Dec.}}{\tan 23^\circ.27'.30''}$;

ϕ has two values, let the least of them, turned into time, $=\phi'$. When the sun's declination is north, the hour which the clock should have marked at the time of the observation is ϕ', if it be before, and $12^h - \phi'$, if it be after the solstice. When the sun's declination is south, $12 + \phi'$ is the hour if it be before, and $24 - \phi'$ if it be after the solstice. The error thus found may either be corrected or allowed for. The clock is afterwards to be regularly compared with the southing of the stars.

20. If a meridian circle pass through any star, the arch of the equator intercepted between that circle and the point of the vernal equinox is called the *Right Ascension* of the star; and if a clock be regulated as above described, the time of a star's passage over the meridian, when turned into degrees, will be equal to its right ascension.

a. Suppose

a. Suppose the hour according to siderial time, when a star passes the meridian, to be $4^{h}. 24^{m}. 50^{sec}.$

$4^{h}.$	=	60°	
$24^{m}.$	=	6°	
$50^{sec}.$	=		12′. 30″
Right Ascen.	=	66°. 12′. 30″.	

b. The preceding observations, for determining the places of the stars, are all supposed to be made on the meridian; and such, when they can be obtained, on account of their simplicity, are preferable to all others. It often happens, however, that the stars must be observed when they are not on the meridian, and their positions, with respect to the immoveable circles of the sphere, must then be derived from spherical trigonometry.

c. The angle which the meridian of a star makes with the meridian of the place of observation, is called the star's *Horary Angle*, as it is the angle which measures the time between the instant of observation and the star's passage over the meridian.

21. Of these five quantities, the Declination, the Altitude, the Azimuth, the Horary Angle of a star, and the Latitude of the place of observation, if any three be given, the other two may be found from the resolution of the spherical

cal triangle PZS, fig. 1. contained by the arches joining the pole, the zenith, and the star.

This general problem contains twenty cases, of which those that follow are the most useful in astronomy.

In all of these, we suppose d to be the declination, a the altitude, h the horary angle, z the azimuth of a star, and l the latitude of the place of observation.

22. Suppose the altitude and azimuth of a star to be observed, and the latitude to be also known; it is required to find the declination, and the horary angle of the star.

Here a, z, l are given, to find d and h, or, in the spherical triangle PZS, the sides ZP, ZS are given, (the complements of l and a) and the angle PZS, between them to find the side PS, and the angle P.

By letting fall a perpendicular from S on the meridian, we may obtain a solution by case 1st and 2d, of oblique-angled spherical triangles.

The solution may also be expressed analytically thus:

Find

Find an angle x, so that

$$\tan x = \cos z \times \cot a, \text{ then}$$

$$\tan h = \frac{\tan z \times \sin x}{\cos (l - x)};$$

$$\text{and } \sin d = \frac{\sin a \times \sin (l - x)}{\cos x}.$$

In applying these formulas it must be observed, that when z, or the azimuth counted from the north, is greater than 90°, cos z, and therefore also tan x, and x itself, are negative.

The latter formula is useful, when the declination of a star is to be determined, and the former when the time is to be determined from the observations of a star not in the meridian. It is more usual, however, when the time is required, to have the declination of the star given, as in the following problem.

23. The declination and the altitude of a star being given, as also the latitude, to find the horary angle and the azimuth.

Here the three sides of the triangle ZPS are given, to find the angles at P and Z.

If d' = complement of Dec.

a' = complement of Alt.

l' = complement of Lat.

sin.

$$\sin\frac{1}{2}h = \sqrt{\frac{\sin\frac{a'+l'-d'}{2}\times\sin\frac{a'-l'+d'}{2}}{\sin l'\times\sin d'}},$$

$$\text{and }\sin\frac{1}{2}z = \sqrt{\frac{\sin\frac{d'+a'-l'}{2}\times\sin\frac{d'-a'+l'}{2}}{\sin l'\times\sin a'}}.$$

By the first formula, the hour can always be determined from astronomical observations, if the time of the star's passing the meridian be known.

The second formula serves to find the meridian, from the observation of the altitude of a known star. The meridian, however, is better found by observing a star when it has the same altitude on the east and west sides of the meridian, and bisecting the difference of the azimuths, as in § 15.

24. Let *a*, *h* and *l* be given, to find *d*; that is, the altitude, the horary angle, and the latitude, to find the declination.

a. Here two sides of the triangle ZPS are given, and the angle opposite to one of them, to find the third side. This falls under Case 7. of oblique-angled spherical triangles.

b. Find

b. Find an arc x, such that $\tan x = \cos h \times \cot l$.

Next find an arc y, so that $\cos y = \frac{\cos x \times \sin a}{\sin l}$;

the declination is the complement of $= y + x$.

This solution is from its nature ambiguous; the sum of y and x must be taken, when the perpendicular from the zenith, on the circle of declination, falls within the triangle; their difference, when it falls without.

25. Let a, d, z be given to find l, or to find the latitude from observing the altitude and azimuth of a star, and knowing also its declination.

Here, again, in the triangle ZPS, two sides are given, and the angle opposite to one of them, to find the third side, the complement of latitude. The perpendicular must be let fall from the star on the meridian; and the distance of this perpendicular, first from the zenith, and then from the pole, is found as in the last case; the sum or difference is the complement of the latitude.

26. Let a, d and h, be given to find l; that is, the altitude and horary angle being observed of a star, of which the declination is known, to find the latitude.

The perpendicular is to be let fall, as in the last case, from the star on the meridian; and there being two sides of the spherical triangle given, and the angle opposite to one of them, the calculation is as before.

This problem is useful for finding the latitude, when two equal altitudes of a star are observed, and the interval of time between the observations. The half of the interval gives the horary angle, and so the latitude may be found as above.

27. In the above formulas it may happen, that $a = 0$, or that the star is in the horizon, or 90° from the zenith. The horary angle is then found, if the latitude and declination are given, from a right angled triangle, of which one of the sides, containing the right angle, is the elevation of the pole, the other, the arch between the star, when rising or setting, and the meridian; and the hypothenuse is the distance of the star from the pole.

In this case, the horary angle, (converted into time), is the time of half the stay of the star above the horizon, (or under it), and if it be called H,

$$\cos H = \tan l \times \tan d.$$

The other side of the triangle, or the azimuth of the rising or setting star, is also called the Amplitude,

and

and is found from this formula,

$$\cos \text{amplit.} = \frac{\sin d}{\cos l}.$$

This is much used in navigation, for ascertaining the variation of the compass.

28. In those stars which are less distant from the pole than the complement of latitude, the triangle formed by the star, the zenith and the pole, becomes right angled at the star, when the circle described by the star, in its diurnal motion, touches the vertical; the azimuth is then a maximum, and is found from the solution of a right-angled triangle, in which the hypothenuse, the complement of latitude, and one side, the complement of declination, are given, to find the angles, one of which is the azimuth, and the other the horary angle of the star.

$$\text{Hence } \sin z = \frac{\cos \text{Dec.}}{\cos \text{Lat.}}$$

$$\text{and } \cos h = \frac{\cot \text{Dec.}}{\cot \text{Lat.}} =$$

$$\cot \text{Dec.} \times \tan \text{Lat.}$$

The first of these gives the azimuth when it is greatest, the second the hour when that happens. They are much used for finding the meridian line.

29. Another

29. Another purpose for which spherical trigonometry is employed, is, when the position of a star relatively to one circle of the sphere is given, to find its position relatively to another which is given in position with respect to the first.

The position of the fixed stars, by the foregoing observations, are found relatively to the equator; that is, the perpendicular arc from the star to the equator, is given, and also the distance of that perpendicular from a given point in the circumference of the equator; but there is another circle, called the Ecliptic, to which it becomes necessary to refer the stars. This circle cuts the equator at the points of the vernal and autumnal equinoxes, and is inclined to it at an angle known from observations which will be afterwards explained, to be nearly 23°. 27′. 30″.

The distance of a star from this circle is called the *Latitude* of the star, and is reckoned north or south according as the star is on the same side of the ecliptic with the North Pole, or the South.

The arc of the ecliptic intercepted between the perpendicular from the star, and the point of the vernal equinox, is called the *Longitude* of the star.

Let the angle formed by the equator and ecliptic, or the *Obliquity* of the Ecliptic be $=\phi$.

30. Suppose

30. Suppose than the right ascension and declination of a star are given, to find its longitude and latitude. Find an arc x, such that,

$$\cot x = \sin z \times \cot \text{Dec.}$$

The declination, if north, is reckoned positive, if south, negative, and x has the same sign with it.

Let $y = x - \phi$, φ being reckoned positive for that half of the ecliptic which is north of the equator, and negative for the opposite, or which comes to the same, positive when the right ascension is less than a semicircle, negative when it is greater:

$$\text{Then } \tan \text{long.} = \frac{\cos y \times \tan \text{Right Asc.}}{\cos x},$$

$$\text{and } \tan \text{Lat.} = \sin \text{Long.} \times \tan y.$$

If tan Long. come out negative, the longitude is greater than a semicircle; if tan Lat. is negative, the latitude is south.

This rule, which is Dr MASKELYNE's, is quite free from ambiguity; it is as simple as the nature of the case will allow, and is deduced from fig. 2., in which A is the intersection of the equator AQ, and the ecliptic AE; S the place of a star, SB an arc of the meridian equal to the declination; AB the right ascension, SD the latitude, and AD the longitude of the star. See VINCE's *Ast.* vol. I. p. 39. LA LANDE computes the angle BAS, and

and the arc AS, in the triangle ABS. Hence the angle SAD being known, and the side AS, in the triangle ADS, the sides AD and DS are found. This is somewhat more prolix. CAGNOLI, § 1448, (2d. edit.) considers the subject differently, but brings out the same solution with Dr MASKELYNE.

DE LAMBRE has computed tables, for faciliating this computation, as it occurs very often in astronomy.

If the star is in the ecliptic, or Lat. = 0, $x = \varphi$, and $y = 0$, therefore $\cos y = 1$, and

$$\tan \text{Long.} = \frac{\tan \text{Right Asc.}}{\cos \varphi}.$$

The changes are obvious, that would make the same formulas apply to the converse of the problem, viz. to find the right ascension and declination from the longitude and latitude being given.

31. By the methods now explained, and chiefly by the observations made on the meridian, catalogues of the stars have been formed, in which their places are all set down in respect of the circles of the sphere, viz. the Ecliptic or the Equator, but most commonly the former.

a. HIPPARCHUS began the first catalogue of the fixed stars, 120 years A. C. "AUSUS," says PLINY, "*rem*

"*rem etiam Deo improbam annumerare stellas, cœlo in*
"*hereditate cunctis relicto.*" Lib. II. cap. 26.

This catalogue was afterwards published by PTOLEMY, with some additions, and contained 1022 stars.

b. The use of the telescope increased the number of the visible stars, as well as the precision with which their places were ascertained. The Britannic Catalogue, published by FLAMSTEAD in 1689, contained nearly 3000 fixed stars. Great additions have been made to this catalogue, particularly by LA CAILLE, who determined the places of nearly 10,000 stars in the southern hemisphere.

Mr WOLLASTON, in 1784, published a catalogue, in which the stars are arranged according to their right ascension and declination.

The different volumes of the *Connaissance des Tems*, contain a catalogue of more than 12,000 fixed stars.

LA LANDE and his nephew observed, in the space of twelve years, 50,000 fixed stars, and determined their places exactly. *Mém. de l'Academie des Sciences*, Paris 1789, 1790. See MONTUCLA, vol. IV. p. 28., &c.

c. Some catalogues, though not very numerous, are highly valued for their accuracy, such as that of LA CAILLE, containing 515 stars; of BRADLEY, containing 309, *Naut. Almanack*, 1773; of MAYER, containing

containing 998; of MASKELYNE, containing only 36; but the places determined with such exactness as to render it extremely valuable. See his *Astronomical Observations*, 1770.

BODE's *Atlas*, Berlin, 1797, is the largest catalogue, reduced into one body, that has appeared; it contains 17,000.

d. The number of the stars, however, visible with the telescope, far exceeds whatever can be reckoned. The bright tract in the heavens, called the *Milky Way*, seems to owe its white appearance to an incredible multitude of stars, which the eye cannot distinguish. Dr HERSCHEL has seen 116,000 stars pass through the field of his telescope in a quarter of an hour; though the field was not more than 15 in diameter.

32. The most obvious distinction among the stars, is founded on their different magnitudes. Those of the first magnitude, are distinguished by particular names; there are only ten visible in Europe, which all astronomers have agreed to belong to that class.

a. The stars visible to the naked eye, are divided, in all, into six classes, not very accurately separated from one another. The ancients counted 15 of the first, 45 of the second, 208 of the third. LA LANDE, § 557.

HALLEY,

HALLEY, *Phil. Trans.* N° 364. Also VINCE, vol. I. p. 505.

b. If *c* be the class reckoned from the first, $13 \times c^2$ is nearly the number of stars in that class.

33. The fixed stars are not scattered over the heavens indiscriminately, but are disposed in groups; to which, from the most remote antiquity, names have been given from certain figures of animals, conceived to be connected with them, which are called *Constellations.*

a. In the beginning of astronomical science, it was only by such a device as this that men could speak of the stars, or describe them to one another. It is a remain of the ancient picture writing, that preceded alphabetical language.

b. The number of the ancient constellations was 48; 24 have been added by the moderns, 14 in the southern hemisphere, and 10 composed out of groups, not included in the ancient arrangement. The stars of each constellation are distinguished by the Greek letters, disposed in reference to their magnitude and position.

For the method of distinguishing the constellations, see LA LANDE, tom. I. § 735., &c.

34. Many of the stars, which, to the naked eye, or through telescopes of small power, appear

pear single, are found, with higher magnifiers, to consist of two, sometimes three or more stars, extremely near to one another.

a. Dr HERSCHEL has observed no less than 700 of these multiple stars, of which only 42 were known before. In some of them the small stars are different in brightness, and in the colour of their light. Thus γ Andromedæ is double; the stars very unequal, the largest a reddish-white, the smallest a sky-blue, inclining to green. See VINCE, § 669.

35. The fixed stars are not entirely exempt from change; several stars which are mentioned by the ancient astronomers having now ceased to be visible, and some being now visible to the naked eye, which are not in the ancient catalogues.

a. It was the appearance of a new star that induced HIPPARCHUS to begin his catalogue. A new star which appeared in Cassiopeia's Chair, in 1572, exceeded Sirius in brightness, and was seen at noon-day. It did not change its place, but gradually decayed, and in about sixteen months disappeared entirely. It is supposed to have appeared before in 945, and 1264. VINCE, § 704.

b. A star of the same kind was seem by KEPLER in 1604; and several similar facts are recorded.

c. Algol,

c. Algol, or β Persei, has been observed to have periodical changes of brightness, that return at the distance of about 2 days and 21 hours. At its greatest brightness, it is of the second, at its least, of the fourth magnitude; the change from the first state to the second is made in about three hours and a half, and the change back again in the same time; during all the rest of the period it preserves its greatest brightness, *Phil. Trans.* 1773. Also VINCE, § 713.

Some other stars are subject to similar variations.

36. Though the fixed stars are without sensible motion relatively to one another, yet many of them, when observed very accurately, are found to change their places slowly.

This was suspected by MAYER, but was first proved by Dr MASKELYNE.

Thus the right ascension of Sirius diminishes annually by two thirds of a second, and his declination, (which is south), increases by 1″.2. Something of the same kind is observed of several others. VINCE, § 724. Whether this motion is always in the same direction, or always at the same rate, is not yet known.

37. In many places of the heavens, spaces faintly luminous, or shining with a pale white light,

light, and of an irregular shape, are discovered; and on applying to them telescopes of great power, they are resolved into a multitude of small stars, distinctly separate, but extremely near to one another. These are called *Nebulæ*.

a. The *Milky Way* is a space of this kind, visible to the naked eye, and encompassing the whole heavens. Dr Herschel's telescope discovers it to consist of a vast multitude of stars, (§ 31. *d*).

Other two *nebulæ* near the South Pole, distinguishable by the naked eye, are called by sailors the *Magellanic Clouds*. The telescope shews them also to be composed of stars.

b. The other *nebulæ* are not visible but with telescopes. Huygens discovered one in Orion's Sword, that appeared a bright spot on a dark ground, and seemed like an opening into some brighter region.

c. Several more *nebulæ* had been observed before Dr Herschel, by Halley, Cassini, La Caille, &c. and a catalogue of 103 was published by the French astronomers, in the *Connaissance des Tems* for 1783. Dr Herschel has given a catalogue of more than 2000 discovered by himself. *Phil. Trans.* 1786, 1789, &c.

38. Dr Herschel has also discovered *nebulous stars*, that is, single stars, surrounded by a faint

faint luminous light. This light has not been resolved into small stars, and Dr Herschel believes it to be the effect of a luminous fluid, *Phil. Trans.* 1791.

39. As the *apparent* motion of a body, in any direction, may arise either from the *real* motion of the body in that direction, or from the motion of the spectator in the opposite; so the apparence of the diurnal motion of the heavenly bodies, round the earth, may either be produced by the real revolution of the heavens from east to west, or by the rotation of the earth, on its axis, from west to east.

This principle is conformable to the experience of every day, and as there are, therefore, two ways of accounting for the phenomena of the diurnal motion, we must choose that which is least liable to objection from other quarters.

40. It is no objection to the supposition, that the diurnal motion of the heavens arises from the motion of the spectator, that he himself is not sensible of his motion. The motion which any body has in common with other bodies, does not affect its state with respect of those bodies, and all the relative motions take place just as if the motion common to them all had no existence.

This

This has been explained under the heads of the first law of motion, and of the distinction between absolute and relative motion.

41. The reality of the diurnal revolution of the heavens, is liable to a great objection, as it supposes that a circular motion, in the same direction, is common to an immense number of bodies, far distant, and entirely detached from one another; and that this motion is so regulated, that their revolutions are all performed in the same time, and in planes parallel to one another.

The revolution of a detached body about a centre, or about an axis, cannot take place without a force constantly acting, to draw it out of the straight line in which it has at every instant a tendency to continue its motion. The revolution of a solid body, like the earth, on its axis, may arise from one original impulse; its continuance requires no new action, but is a consequence of the *inertia* of matter.

It was to obviate the difficulty arising from the detached and distant situation of the bodies, to which motions so perfectly coincident were ascribed, that the hypothesis of crystalline orbs was invented. To those who do not admit the existence of these orbs, the diurnal revolution of the heavens can have no probability.

42. The

42. The physical and mechanical objections to the rotation of the earth being entirely obviated, while many press so hard upon the opposite hypothesis, the simplicity of the explanations afforded by the former, justifies us in admitting it as the cause of the apparent diurnal revolution of the heavens, at least till some fact, or some principle inconsistent with it shall be discovered.

SECT. II.

OF THE ATMOSPHERICAL REFRACTION.

43. A RAY of light, in passing through the atmosphere, is bent into a curve, in the same vertical plane with the original ray, and concave toward the surface of the earth. But the object from which the ray comes, is seen in the direction which the ray has when it enters the eye, and therefore it appears elevated above its true place. This is called the *Atmospherical*, and sometimes the *Astronomical Refraction.*

a. The effect of the atmospheric refraction alters the place of an object only in a vertical plane; it increases

creases the altitude, but does not affect the azimuth.

b. Hence all the altitudes, measured as in the preceding sections, require corrections to be applied to them before the true altitudes are obtained.

The method of making these corrections is now to be explained.

44. From the principles of optics, it is known, that the rays which pass through the strata of the atmosphere at right angles, or which come from stars in the zenith, suffer no refraction, and that, at all other elevations, the quantity of the refraction is nearly as the tangent of the zenith distance.

a. If x be the true distance of a star from the zenith, and y the refraction, so that the apparent distance is $x-y$, then, by the nature of refraction, $\sin x$ is to $\sin(x-y)$ in a constant ratio, suppose that of m to n; and since $\sin(x-y)=\sin x \,.\, \cos y$ $-\cos x \,.\, \sin y$, $\frac{n}{m} \sin x = \sin x \,.\, \cos y - \cos x \,.\, \sin y$, or $\frac{n}{m} = \cos y - \frac{\cos x}{\sin x} \,.\, \sin y$. Now if y be very small, $\cos y = 1$ nearly; and therefore $\sin y = \left(1-\frac{n}{m}\right)\frac{\sin x}{\cos x} = \left(1-\frac{n}{m}\right) \tan x$. As $\sin y = y$ nearly

nearly, $y=\left(1-\frac{n}{m}\right)\tan x$; where the constant coefficient $1-\frac{n}{m}$ is to be determined by observation.

6. It is evident that this approximation is imperfect, because it gives y infinite when $x=90°$, or when the star is in the horizon; and in all cases when the altitude is small, it gives the refraction too great. Dr BRADLEY found that the refraction was more nearly as the tangent of the apparent zenith distance, diminished by a certain multiple of the refraction, or that $y=A.\tan(x-b\,y)$. He also found $A=57''$ and $b=3$ nearly, so that $y=57''\tan(x-3\,y)$.

To use this formula, we may suppose, for a first approximation, $y=57''\tan x$, when x is not very great.

If $x=90°$, we must assume y, and correct the value by trials.

The manner in which the constant quantities A and b have been determined, remains to be explained.

45. As the determination of the latitude of a place, and of the declination of a star, involve the effect of refraction, they are not to be employed for finding the refraction, except when

a considerable error in them will not produce any sensible error in the refraction to be found.

In conformity with this rule, the following method of finding the horizontal refraction may be employed.

46. A star which rises and sets due east and west, would be as long under the horizon, as it is above it, if there were no refraction. It will be found, however, by observation, to be longer above than below the horizon: take half the difference, and reduce it to degrees, or parts of a degree; then multiply this last by the cosine of the latitude of the place; the product is the horizontal refraction.

a. In this way, the mean horizontal refraction will be found to be 33′; and here it would require a great error in the latitude to produce an error at all sensible in the refraction.

b. This method requires a situation, where the stars to be observed rise and set in the sea. It may, however, be extended, though with less simplicity, to all the stars, and to their descending, not below the horizon, but below a given altitude, or a given parallel to the horizon.

47. Observe the altitude and azimuth of a star of a known declination at the same instant; from

from the azimuth, the polar distance and the complement of latitude, compute the altitude; the difference between this and the observed altitude is the refraction.

a. This may be easily done with the Astronomical Circle, which measures altitudes and azimuths at the same time. The circumpolar stars are well suited to this observation.

The triangle may be resolved by NAPIER's rules, *Mauduit*, p. 76. anot. 2. See other methods, VINCE, § 176., &c. WOODHOUSE, p. 85., &c. LA LANDE, 2170., &c.

b. If the time from the altitude of a known star, on the one side of the meridian, to the same altitude on the other side, be observed, the apparent altitude may be calculated from thence; and if a circumpolar star is used, great accuracy may be obtained in this way, as a small change in the altitude produces a great one in the horary angle; so that to observe this latter angle, is the most exact way of coming to the knowledge of the apparent altitude.

48. The refraction varies with the state of the barometer and thermometer, and, for the same altitude, is nearly proportional to the density of the air at the earth's surface.

a. Dr

a. Dr BRADLEY has given a formula, expressing this variation of the refraction.

Let b = height of the mercury in the barometer in inches, h = height of FAHRENHEIT's thermometer, z = the zenith distance, r = the mean refraction computed by the rule, (§ 44. *b.*), the correct refraction $r' = \frac{b\,r}{29.6} \times \frac{400}{350+h}$.

This formula would be improved, by introducing the expression for the density or specific gravity of the air from vol. I. § 333. It will then be

$$r' = \frac{br}{29.6} \times (1 - 0.00222\,(h - 50)).$$

49. Not only are the stars elevated by refraction, but all terrestrial objects are elevated from the same cause, by an angle equal to that which the straight line drawn from the eye to the object makes with the tangent to the path of the ray at the point where it enters the eye.

The refraction of the heavenly and terrestrial bodies is differently estimated. That of the celestial body is the angle contained between the tangent to the curvilineal path of the ray where it is first acted on by the atmosphere, and the tangent to the same curve when it enters the eye. The refraction of a terrestrial body is the angle contained between the tangent at the eye and the chord of the arch intercepted between the object and the eye. Near

Near the earth's surface, the curvilineal path of the ray of light may be supposed to coincide with the circle of equal curvature.

50. If the elevation of the top of a mountain from a point in the plane below, and the depression of that point from the top of the mountain, be both observed at the same time, the angle subtended at the earth's centre by the distance between these points, added to the observed elevation, and the sum diminished by the depression, is double of the refraction.

This supposes the path of a ray of light, for a small part, to coincide with a circle. If, in fig 4., the arch from B to A be the path of a ray, and if AH and BF be perpendicular to AC, BC, in A and B; the tangents EA, EB being drawn to the path of the ray, HAE is the apparent elevation of B from A, and FBE the apparent depression of A from B; the true elevation being HAB, and the true depression FBA. It is evident that FBA = BAH+ACB, or true Dep. = true Elev.+Hor. Ang.; that is, true Elev. = true Dep.—Hor. Ang. But true Elev. = app. Elev.—Ref.; and true Dep. = app. Dep.+Ref.; therefore 2 Ref. = Hor. Ang.+app. Elev.—app. Dep.

51. The

51. The terrestrial refraction found by means of the preceding theorem, when the elevation is not very great, varies from $\frac{1}{4}$ to $\frac{1}{24}$ of the angle subtended by the horizontal distance of the objects; and the radius of curvature of the ray, therefore, varies from twice to twelve times the radius of the earth.

In the mean state of the atmosphere, the refraction is about $\frac{1}{14}$ of the horizontal angle, and the radius of curvature of the ray seven times the radius of the earth.

The terrestrial refraction must vary with the density of the air, or with the barometer and thermometer. The great differences, however, remarked above, must be owing to some other cause.

In the measurement of heights, the angle of elevation should be diminished by one-fourteenth of the angle corresponding to the horizontal distance, supposing the refraction to be of the mean quantity.

52. The effect of refraction may also be allowed for, by computing the correction of curvature, as in § 248. vol. I., and taking one-seventh of it, for the number of feet, by which the object is rendered by the refraction higher than it ought to be.

If

If L be the length of the horizontal line in English miles, the correction for curvature in feet is $\frac{2L^2}{3}$,

and for refraction $\frac{2L^2}{21}$,

and this last is to be subtracted from the former, which gives $\frac{12L^2}{21}$ or $\frac{4L^2}{7}$ for the quantity to be added to the height as calculated trigonometrically, including both the corrections for curvature, and, in the mean state of the atmosphere, for refraction.

When a height is calculated trigonometrically from the angle of apparent elevation observed at a given distance, both the preceding corrections may be made on the angle itself, which on some occasions is more convenient. Divide the horizontal distance in feet by 6075, the feet in a geographical mile or a minute of a great circle of the earth. Let the quotient $= H$, and let the observed angle of elevation $= E$, then the true angle corrected both for curvature and refraction is $E + 0.428\ H$.

SECT. III.

FIGURE OF THE EARTH.

53. THE figure of the earth is understood to be determined by a surface at every point perpendicular

dicular to the direction of gravity, or to the direction of the plumb-line, (§ 24. *b*.).

This surface is the same that the sea would have if it were continued all round the earth; or, if we were to trace curve lines, by levelling from a given point round the earth, in every direction, till they returned into themselves, the superficies in which all these lines would lie, is that which we consider as the superficies of the earth. The given point may be supposed any one, on the level of the sea.

The figure bounded by this superficies, is that which is really measured by the combined methods of astronomy and practical geometry, and is to be carefully distinguished from the *actual* figure of the earth, including all its inequalities; or from an average figure that should leave out as much of solid matter above it, as is included of empty space under it.

54. The length of an arch of the meridian, traced on the superficies above defined, may be measured by observing the latitude of the two extremities of the arc, and then measuring the distance between these points in fathoms, toises, or any other known measure.

The distance, as measured on the surface, divided by the degrees, and parts of a degree contained in the difference

difference of the latitudes, will give the length of a degree.

Eratosthenes was the first who applied this method to the estimation of the earth's circumference. By measuring, or rather estimating, the difference of latitude between Alexandria and Syene, and also their distance, he concluded the circumference of the earth to be 250000 stadia. Montucla, *Histoire des Math.* tom. I. p. 242. 2d edit.

The length of a degree, or of any arch of the meridian, is determined by taking two points nearly at the distance of the arch required, and nearly north and south of one another. A series of triangles is then to be carried from the one point to the other by means of stations taken on the tops of hills or other elevated grounds. The angles of these triangles are to be measured, as also the azimuths of the sides, at the points where the series of triangles begins and ends. Thus, there is a series of triangles all given in *species* which connects the two extreme points, and the bearings of the sides of these triangles, in respect of the meridian of the first station, are also given.

The lengths of the sides of the triangles in known measures, toises or fathoms, is next found by measuring a base on a level ground, and connecting it by angles with the sides of one of the triangles. In all this we proceed as if the triangles were plane, whereas they are in fact spherical, and the three angles

angles of each exceed two right angles, by a quantity called the *Spherical Excess*; which is to four right angles, as the area of the triangle to half the superficies of the globe. By knowing the spherical excess, the errors in the measures of the angles are discovered; and it is also known, that if from each of the angles of any of the triangles, be subtracted one-third of the spherical excess, the sines of the angles so corrected are proportional to the lengths of the sides; so that all the triangles, though strictly speaking spherical, may be resolved by the rules of plane trigonometry.

In this way, the distance of the two points at first assumed is computed, as also their distance reduced to the meridian of either of them.

55. When, by such accurate observations, the lengths of degrees were determined in different latitudes, they were found to increase gradually from the Equator to the Pole.

The radius of curvature of the meridian, therefore, increases as we go toward the Pole, and the curvature itself diminishes. The earth, therefore, is not a sphere, but is flattened at the Poles, so that the axis from Pole to Pole is less than the diameter of the Equator. Though it is only by experiment that the true figure of the meridian can be discovered, it has been found necessary to assume hypothetically, for its figure, the curve which is next in simplicity

simplicity to the circle, viz. the Ellipsis, and also to suppose the superficies of the earth to be that of a spheroid generated by the revolution of this ellipsis, about its shorter axis.

In many complex cases, this mode of approximating to the truth, by probable assumptions, is the simplest that can be pursued. The hypothesis thus assumed, must be rigorously submitted to the test of experience.

56. The solid contained by the radius of curvature, at any point in an ellipsis, and the square of the semiparameter of the greater axis, is equal to the cube of the normal at the same point.

That is, if a and b are the semiaxes, r the radius of curvature, n the normal at any point, $n^3 = \frac{b^4}{a^2} \times r$.

See FRISIUS *de Analysi Sect. Con.* *Opera*, tom. I. p. 96. prob. 32. NEWTON's *Conic Sections*, prop. 78. cor. 2.

57. Hence the radius of curvature, at any point of the meridian, and consequently the length of a degree at that point, may be expressed in terms of the latitude: if r be the radius of curvature at a point, of which the latitude is λ, a and b denoting as before,

$$r = \frac{a^2 b^2}{(a^2 \cos\lambda^2 + b^2 \sin\lambda^2)^{\frac{3}{2}}}.$$

Let

Let ADB be one half of the meridian, A and B points in the Equator, C the centre of the earth, D the Pole, EA a perpendicular to the meridian at E, a point of which the latitude is $\lambda = \text{EGA}$, H the centre of curvature, F a perpendicular on the axis; then EG is the normal, or n, GF the subnormal $= s$; let CF $= x$, and FE $= y$; then $y^2 = \frac{b^2}{a^2}(a^2 - x^2)$. But, by the property of the subtangents of the ellipsis $x = \frac{a^2}{b^2} s$; also $s = n \cos \lambda$, and $y = n \sin \lambda$; therefore, by substitution,

$$n^2 (b^2 \sin \lambda^2 + a^2 \cos \lambda^2) = b^4,$$

$$\text{and } n = \frac{b^2}{(a^2 \cos \lambda^2 + b^2 \sin \lambda^2)^{\frac{1}{2}}}.$$

Now, by § 56. $r = \frac{a^2}{b^4} n^3$; therefore,

$$r = \frac{a^2 b^2}{(a^2 \cos \lambda^2 + b^2 \sin \lambda^2)^{\frac{3}{2}}}$$

If D be the length of a degree in lat. λ, and $m = 57^\circ.2957795$, the number of degrees in an arc equal to the radius, then $r = m$ D,

$$\text{and D} = \frac{a^2 b^2}{m (a^2 \cos \lambda^2 + b^2 \sin \lambda^2)^{\frac{3}{2}}}.$$

58. In

58. In an ellipse where the eccentricity is small, or where a and b differ but by a small quantity c, this general formula may be reduced to more simplicity, by extracting the root of the denominator, and rejecting the powers of b greater than the first; we have then

$$m\,D = a\left(1 - \frac{2c}{a} + \frac{3c}{a}\sin \lambda^2\right).$$

a. This value of m D may be changed into another, more convenient in calculation, by substituting for $\sin \lambda^2$ its value, $\frac{1 - \cos 2\lambda}{2}$, from which is obtained

$$m\,D = a\left(1 - \frac{c}{2a} - \frac{3c}{2a}\cos 2\lambda\right).$$

b. At the Equator, $\lambda = 0$, and $\cos 2\lambda = 1$; so that

$$m\,D = a\left(1 - \frac{2c}{a}\right) = a - 2c.$$

c. At the Pole, $\lambda = 90°$, $2\lambda = 180°$; and since $\cos 180° = -1$, $m\,D = a + c$.

The degree of the meridian at the equator, is therefore to the degree at the pole as $a - 2c$ to $a + c$.

d. In the parallel of 45°, $2\lambda = 90°$, and $\cos 2\lambda = 0$; therefore $m\,D = a - \frac{c}{2}$.

The

The radius of curvature at the parallel of 45°, or m D, is nearly an arithmetical mean between the radius of the equator and half the earth's axis.

The degree in the parallel of 45 is also an arithmetical mean between any two degrees equally distant from it on the north and on the south.

e. The degree in any latitude is understood to be that of which the middle point is in that latitude.

f. If, therefore, D be found by actual measurement in any two known latitudes, we shall have two equations, in which a and c are the only unknown quantities, and from which, therefore, they may be determined.

59. The lengths of two degrees, of which the middle points are in given latitudes, being known; it is required from thence to determine the diameter of the equator, and the axis of the earth, that is, the longer and the shorter axis of the elliptic meridian.

a. Let D and D′ be the given degrees, (the least, or that nearest the equator, being D), λ and λ' the latitudes of their middle points, a the semitransverse axis of the meridian, c the difference of the semiaxis; we have the equations,

m D =

$$m\,\mathrm{D} = a - \frac{c}{2} - \frac{3c}{2}\cos 2\lambda, \text{ and}$$

$$m\,\mathrm{D}' = a - \frac{c}{2} - \frac{3c}{2}\cos 2\lambda'.$$

$$\text{Hence } c = \frac{2m(\mathrm{D}' - \mathrm{D})}{3(\cos 2\lambda - \cos 2\lambda')},$$

$$\text{and } \frac{c}{a} = \frac{2(\mathrm{D}' - \mathrm{D})}{3\,\mathrm{D}(\cos 2\lambda - \cos 2\lambda')}.$$

Also, if $\frac{c}{a} = n$,

$$a = \frac{m\,\mathrm{D}}{1 - \frac{n}{2} - \frac{3n}{2}\cos 2\lambda}.$$

b. These formulas will be reduced to others, more convenient for logarithmical calculation, by substituting for $\cos 2\lambda - \cos 2\lambda'$, its value, viz.

$$2\sin(\lambda + \lambda') \times \sin(\lambda' - \lambda);$$

$$\text{then } c = \frac{m(\mathrm{D}' - \mathrm{D})}{3\sin(\lambda' + \lambda) \times \sin(\lambda' - \lambda)};$$

$$\text{and } \frac{c}{a} = \frac{\mathrm{D}' - \mathrm{D}}{3\,\mathrm{D}\sin(\lambda' + \lambda) \times \sin(\lambda' - \lambda)}.$$

When λ is nothing, or when one of the degrees is at the equator, $\lambda + \lambda'$ and $\lambda' - \lambda$, are each equal to λ', so $c = \frac{m(\mathrm{D}' - \mathrm{D})}{3\sin^2\lambda}$; therefore the excess of the

degree,

degree, in any latitude, above the degree at the equator, when divided by the square of the sine of the latitude, should always give the same quotient; or the excess of the degrees of the meridian above the degree at the equator, should be as the squares of the sines of the latitudes.

$$\text{Since } c = \frac{m(D' - D)}{3 \sin(\lambda' + \lambda) \times \sin(\lambda' - \lambda)},$$

$$D' - D = \frac{3c}{m} \sin(\lambda' + \lambda) \times \sin(\lambda' - \lambda).$$

c. If, then, D′ and D are two contiguous degrees, so

$$\text{that } \lambda' = \lambda + 1°, D' - D = \frac{3c}{m} \sin(2\lambda + 1°) \times \sin 1°;$$

and since $\sin 1° = .01745$, $D' - D =$

$$\frac{3c \times .01745}{m} \sin(2\lambda + 1°).$$

The contiguous degrees, therefore, differ by a quantity proportional to the sine of twice the middle latitude. The difference is a maximum, when $2\lambda + 1 = 90°$, or when the middle latitude is 45°.

The

The quantity $\frac{c}{a}$ is called the Compression, and determines the species of the ellipsis.

d. We shall now take for the determination of the figure and magnitude of the earth, the five arches subjoined, as those that have been measured with the greatest care, and the best instruments; as being the largest also, and the most distant from one another.

	Lat.		Deg. in toises.	Deg. in fathoms.	Country.
I.	0°	.0′.0″	56749	60480.2	Peru.
II.	11	.0.0	56755	60486.6	India.
III.	45	.0.0	57011	60759.4	France.
IV.	52	.2.2	57074	60826.6	England.
V.	66	.20.10	57192	60952.4	Lapland.

As five quantities may be combined, two and two, in ten different ways, ten results may be deduced from the comparison of these degrees; and if the meridian were truly elliptical, and if there were no error in the observations, all these results would coincide. As the latter supposition cannot be expected to hold, we must look for some difference in the results, and must choose only those combinations, in which the degrees are considerably distant from one another, because in that way the errors of observation will least affect the conclusions. Such are the 1st and 3d, 1st and 4th, 1st and 5th; 2d and 3d, 2d and 4th, 2d and 5th; 3d and 5th.

e. The 3d degree, or that which is bisected by the parallel of 45°, is to be reckoned the most accurate of all, as being deduced from the actual measurement of an arch of more than 12 degrees. This arch belongs to the meridian of Paris, which has been continued by the French mathematicians north to Dunkirk, lat. 51° 2′ 9″ N., and south to Formentera, the southernmost of the Balearic isles, in lat. 38° 38′ 56″, the distance being 705188.77 toises, (*Base Metrique*, tom. III. p. 298.) The arch of the same meridian from Dunkirk, till it is intersected by the parallel of Greenwich Observatory, lat. 51° 28′ 39″½, was also measured by General Roy, and found to be 25238.5, or, more exactly, by De Lambre, on applying certain corrections, 25241.9 toises. Thus the amplitude of the arch between the parallels of Greenwich and Formentera, is 12° 48′ 43″.5, and the distance 730430.67. This arch, however, is not exactly bisected by the parallel of 45°, Greenwich being 6° 28′ 39″.5 distant from that parallel, and Formentera only 6° 20′ 3″.99. To have the length of an arch beginning at Formentera, and terminating just as far to the north, as Formentera is to the south of 45°, we must subtract from the distance of Greenwich the number of toises corresponding to 8′ 35″.51, which, as may be inferred from the measurement itself, is 8165.88 toises; and thus we have 722264.79 for the length of an arch extending 6° 20′ 3″.99 on each side of the parallel of 45°.

Now,

Now, it was observed (§ 58. *d.*) that the degree in lat. 45° is an arithmetical mean between any two degrees equally distant from it no the north and south; and therefore if an arch of any number of degrees, for example 6, be measured on the south of 45°, and an arch of as many degrees on the north, the two together will be 12 times the degree in the parallel of 45°. If, therefore, the above extent, 722264.79, be divided by 12.6689, the amplitude of the arch, the quotient 57011t is the degree of the meridian bisected by the parallel of 45°.

f. The degree at the Equator is deduced from an arch of 3° 7′ measured in Peru by the French and Spanish academicians. It is stated by BOUGUER at 56753t; by CONDAMINE, at 56749; and we prefer the latter number, as there is reason to think, from comparing that degree with others, that it is too great. *Figure de la Terre*, par M. BOUGUER, Sect. v. p. 272.

g. The degree lat. 11°, is a mean of 6 degrees, measured by Major LAMBTON in Indostan. *Asiatic Researches*, vol. XII. p. 94.

h. The degree in lat. 52° 2′ 20″ is from an arch of the meridian measured by Colonel MUDGE in the south of England. The mean is given here from a paper, *Phil. Trans.* for 1812, p. 332.

i. The degree at the Polar Circle, or in lat. 66° 20′ 10″, is from an arch of the meridian lately measured by SWANBERG and other Swedish academicians. *Exposition*

position des Operations faites en Lapponie, &c. par SWANBERG. Stockholm, 1805.

k. Of all these degrees, it may be said, that an error of 30 toises in the length, or 2″ in the amplitude of the arch, is more than can be reasonably supposed. If the hypothesis of an elliptic meridian agree with them nearer than this quantity, it must be considered as having the support of observation. But if it is found that these arches cannot be reconciled with the elliptic hypothesis, without supposing greater errors than those just mentioned, that hypothesis must be either rejected, or regarded as doubtful.

60. The five degrees in the last article, coincide in giving very nearly the same compression to the earth at the poles, and may all be represented by the same equation, to an exactness much within the limits that have been assigned (§ 59. *k.*)

a. By combining the degrees in the seven ways mentioned, it will be found, that $\frac{c}{a}$ is between .00315 and .00325.

b. The mean between these, or .0032, it that which, on the whole, seems the nearest to the truth. It makes the sum of all the errors in the five degrees amount only to 21 toises, taking them with the same sign. Taking them with their proper signs, they nearly

nearly destroy one another. The compression may therefore be stated at .0032 $= \frac{1}{312.5}$; and the equation which does most nearly represent the degrees of the meridian, will from thence come out,

$$D = 57011^{t} - 272^{t}.65 \cos 2 \lambda.$$

In fathoms,

$$D = 60759.472 - 290.576 \cos 2 \lambda.$$

In miles, $D = 69.044 - .3299 \cos 2 \lambda.$

c. Hence, by the formulas (§ 59. *b.*),

	Toises.	Fathoms.	Miles.
$c =$	10469.58 =	11158.8 =	12.680
$a =$	3271743.00 =	3486858.8 =	3962.349
$b =$	3261273.42 =	3475700. =	3949.669

Radius of curvature for the parallel of 45° =

$$a - \frac{c}{2} = 3266508^{t}.21 = 3481279^{f}.4 = 3956.009$$

miles. The miles meant here are English miles.

d. The circumference of the Elliptic Meridian may be found nearly by multiplying the mean degree, or that in the parallel of 45°. by 360. The result is 24855.84 miles.

The

The circumference of the Equator is 24896.16 miles, a little more than 40 miles greater than the preceding.

The circumference of the Meridian may be found more accurately by the theorem for the rectification of the ellipsis. See LA CAILLE, *Leçons Elementaires de Math.* § 954. Also *Base Metrique*, tom. II. p. 676.

The French, from their late measurement, compared with that in Peru, make the compression .00324, and the quadrant of the meridian 5131111 toises; which gives for the entire circumference 5468481.54 fathoms, or 24856.72 miles; about 1 mile greater than the result obtained above.

e. The Geographical Mile, or that of which there are 60 in the length of the mean degree, is 1012.6 fathoms, = 6075.6 feet.

61. The semidiameter of the earth belonging to any latitude λ, is nearly equal to $a\ (1 - \frac{c}{a}\sin^2\lambda)$.

a. This formula is found, by expressing the semidiameter of the elliptic meridian, in terms of the latitude, in a manner similar to that employed

§ 57,

§ 57, 58., then extracting the square root, and rejecting the powers of the compression greater than the first.

See CAGNOLI *Trigonometrie*, § 1558.

b. The tangent of the angle ACE (fig. 3.), which the diameter answering to the latitude λ, makes with the greater axis of the meridian, or with the plane of the equator, is $\frac{b^2}{a^2} \times \tan \lambda$. CAGNOLI, § 1339.

c. The angle at the centre being thus found, the angle which the semidiameter CE makes with the vertical EG is also found, being the difference between the former angle and the latitude of the place. The angle which the diameter makes with the vertical is greatest at 45°, where it amounts to 11′ 9″ when the compression is .00324. *Base Metrique*, tom. III. p. 292. If the compression is greater, this angle also becomes greater.

d. The figure of the earth may not only be determined by comparing two degrees of the meridian with one another, but it may also be found by comparing a degree of the meridian, with the degree of a great circle perpendicular to the meridian, in the same latitude.

62. If we suppose the earth to be cut at any point by a plane perpendicular to the meridian in that

that point, the centre of curvature of this section, at the point where it cuts the meridian, is the point in which the direction of gravity, or of the plumb-line, intersects the axis of the earth.

a. The direction of gravity, if the earth be a solid of revolution, passes always through the axis of the earth. If, therefore, we conceive the plumb-line to be carried over an indefinitely small arch of the perpendicular to the meridian, either to the east or west, its direction will intersect the axis at the same point where it intersected it before, which point, therefore, is the centre of curvature of the arch, or the same with K (fig. 3.). EK is greater than EH, and the degree of the perpendicular arch is greater than the degree of the meridian in the same ratio.

b. The radius of curvature of the arch perpendicular to the meridian, is therefore the *normal* of the meridian, relatively to its shorter axis, or it is

$$\frac{a^2}{\sqrt{a^2 \cos^2 \lambda + b^2 \sin^2 \lambda}}.$$ *Edin. Trans.* vol. v. p. 14.

63. If D be the degree of the meridian, at a point of which the latitude is λ, and Δ the degree of the curve perpendicular to the meridian at the same point,

$c =$

$$c = \frac{m}{2} (\Delta - D) \times \frac{1}{\cos^2 \lambda};$$

$$a = m\,\Delta - \frac{m}{2} (\Delta - D) \tan^2 \lambda;$$

$$\frac{c}{a} = \frac{\Delta - D}{2\,\Delta \cos \lambda^2} \text{ nearly.}$$

Thus the figure of the earth is determined by the degree of the meridian in any latitude, compared with the degree perpendicular to it.

The degree of the section perpendicular to the meridian, is to the degree of the circle parallel to the equator, that is, to the length of the degree of longitude, as 1 to cos λ.

The degree of longitude is therefore $\Delta \cos \lambda$. *Edin. Trans.* vol. v. p. 26.

The manner in which the amplitude of the celestial arch is measured in the case of a perpendicular to the meridian, is not so direct as that which is followed in the case of an arch of the meridian itself. It is best done by azimuths or by determining the *convergency of the meridians.* If P be the pole, (fig. 5.), AP an arch of the meridian, AD an arch at right angles to it, or nearly so: Let the latitude of A, and also the angle PAD be found, and again at D the angle PDA; then in the spherical triangle PAD, the angles at A and D are given, and also the side AP, from which the arch AD may be computed in degrees and minutes; and its length also having been measured,

measured, the length of a degree becomes known. It is not necessary that AD should be accurately at right angles to either of the meridians AP, DP: If it is nearly so, its length can easily be reduced to that of the perpendicular.

This method of finding the amplitude of an arch perpendicular to the meridian, is only applicable in high latitudes, where the convergency of the meridians is considerable. Near the equator, where the meridians became almost parallel, a small error in determining the azimuths will produce a very great one in the amplitude of the arch, so that this method cannot be safely employed. The manner of finding the amplitude of the perpendicular arch accurately in such cases, depends on the methods of finding the difference of longitude, which are to be explained in the next section.

It appears from the preceding investigations, that the earth is an *oblate spheroid*, generated by the revolution of an ellipsis about its shorter axis, that axis being to the longer axis as .9968 to 1. All the arches, however, that have been measured, do not agree equally in bringing out this result. In general, though the conclusions from arches, which are large, and at a considerable distance, are consistent with one another, the contrary holds where arches very near to one another, and more especially, contiguous portions of the same arch, are compared together. It has been shown, that, according to the elliptic hypothesis, the differences of the contiguous

contiguous degrees ought to be proportional to the sines of twice the middle latitude, (§ 59. *c*.), and therefore ought to follow a very regular progression. This, however, has not been found to take place in any instance of actual measurement. In the great arch measured in France, in those of the Peninsula of India, and in the south of England, the contiguous degrees have differed very irregularly, and by such considerable quantities, as could not be explained by any probable error of observation. Local irregularities in the figure of the earth, manifesting themselves by the deflection of the plumb-line, seem to give rise to these anomalies. The deflections are but small, and disappear altogether when arches of great extent, and differing by large quantities, are compared with one another.

This subject will be more fully considered under the head of Physical Astronomy.

SECT. IV.

GEOGRAPHICAL PROBLEMS.

64. The situation of a point in a given superficies is determined, when its distances are known from

from two planes, which are given in position with respect to that superficies.

a. That the determinations thus afforded may be the simplest possible, the two planes ought to be at right angles to one another; and if the superficies is one having a centre, the planes should pass through that centre.

b. In the case of the earth, the plane of the Equator having its position fixed by the diurnal motion, is naturally pointed out as one of the fixed planes, to which the positions of places in the earth's surface are to be referred.

The position of every place, relatively to the equator, is determined by finding its latitude as above defined.

c. The other circle to which the position of places on the earth's surface is to be referred, must necessarily be a Meridian (*a*); but as none of the meridians is distinguished from another, by any circumstance in the diurnal motion, of which they all partake alike, the particular meridian that is to be fixed on for the determination of geographical positions, is a matter of arbitrary arrangement.

d. When a meridian is chosen for a first meridian, or that to which all positions are to be referred, it is not by directly measuring the distance from it that such a reference is made, but by measuring the angle which the plane of the meridian passing through

through any given place, makes with the plane of the first meridian. This angle is called the *longitude* of the place, and the diurnal motion furnishes us with the means of determining it. It is measured by the arch of the equator, intercepted between the first meridian and the meridian of the place, and is reckoned east or west, according as the place is east or west of the first meridian.

The ancients took for their first meridian, or that from which their longitude was counted, the meridian of the Fortunate Isles, a line passing, as they conceived, through the western extremity of the habitable earth. Many of the moderns have employed the same meridian, or rather that of the Island of Ferro, one of the most westerly of the Canaries. In general, however, nations employ the meridian of their own metropolis, or of their principal observatory; as we do that of Greenwich, the French that of Paris, &c. It has been proposed to take the meridian of Mont Blanc as the first meridian, being that of a point very remarkable in the natural history of the globe. It would be inconvenient, however, to take for a first meridian any point where astronomical observations are not constantly made.

65. The hour, as reckoned under any two meridians, is different, and the difference is proportional to the difference of longitude, or the angle which

which the planes of these meridians make with one another.

The hour at any place, as we have already seen, is determined by the passage of a certain star, or a certain point in the heavens, over the meridian of that place. But the star comes to any meridian sooner than to another farther to the west, by a space of time which is to 24 hours as the angle made by the two meridians to 360 degrees. For every 15°, therefore, contained in the angle which the meridians make, or in the difference of longitude, one hour is to be reckoned, by which the account of time at the more westerly place is later, or the clock slower than at the other. For all other angles, the proportion is the same; for one degree, four minutes of time; for one minute of a degree, four seconds of time, &c.

66. If, therefore, we could find under one meridian the time which they reckon at the same instant under another, we should have the difference of longitude, by converting the difference of time into degrees at the rate just mentioned.

a. An obvious way, therefore, of discovering the difference of longitude of any two places, is to have a watch, or portable *chronometer*, well regulated, according to the time at one of the places, and then to carry it to the other; where, on being compared with

with the time, as reckoned there, it will give the difference of longitude.

b. Suppose at the first place, by repeated observations, the star Sirius was found to pass the meridian at 10^{h} 24^{m} 35^{sec} by the watch, and at the other at 8^{h} 16^{m} 22^{sec} by the same watch; the difference, 2^{h} 8^{m} 13^{sec}, converted into degrees, gives 32° 3′ 15″ for the difference of longitude, by which the first of the two places is east of the second.

67. In like manner, any phenomenon, the beginning or end of which is seen at the same instant by observers under different meridians, affords the means of determining the difference of longitude.

For, by this means, the difference of the reckoning at the two places is ascertained just as by the chronometer.

Many of the phenomena of the heavens, as will be afterwards explained, serve as signals of this kind.

Such signals may also be given on the earth, by the sudden kindling or extinguishing of lights on some elevated station, from which they may be seen at considerable distances.

The observation of the convergency of the meridians, is also another way of determining the longitude, as already observed. To it, and the two now mentioned,

tioned, all the methods of finding the longitude may be reduced.

68. If the latitudes of any two places are given, and also their difference of longitude, their distance may be found by spherical trigonometry.

a. If the earth is considered as a sphere, then, in the spherical triangle contained by the arches joining the two places with one another, and with the pole, two sides are given, viz. the distances from the pole, or the complements of latitude, and the angle at the pole, or the difference of longitude; and therefore the 3d side may be found by the 2d case of oblique-angled spherical triangles. This side is the distance of the places expressed in degrees, &c.; and may be turned into miles, by multiplying by 69.044, the mean length of a degree, (§ 60. *b.*)

If the angles at the base or the azimuths are also required, it will be best to resolve the triangle by NAPIER's Formula. See *Elem. of Geomet.* Edin. 1810, p. 378. See also WOODHOUSE's *Trigonometry*, p. 126.

b. But if the spheroidal figure is to be taken into account, the calculation becomes more complex. For as, on this supposition, the directions of the plummets AD, BF, (fig. 6.) at the two places, if their latitudes are different, do not meet the axis in the same point, these three lines do not contain a solid

 angle,

angle, and therefore the rules of trigonometry cannot be directly applied to them. If, however, C be the centre of the spheroid, and if AC and BC be joined, the angles PCA, PCB, are deduced from the latitudes, § 61. *b.* Then, in the solid angle at C, are given the two plane angles PCA, PCB, and the inclination of their planes, viz. the difference of longitude, or the angle at P; therefore the angle ACB may be found by the same case of spherical triangles as before. Hence the straight line AB is also found, the radii CA, CB being given, § 61.

c. In this way also, are found the angles at the base of the triangle PAB, or those which the plane ACB makes with the planes ACP, BCP. These, however, are not the true azimuths, which are the angles that the plane ADB makes with ADP, and that ABE makes with PEB.

To find these last; if DB be drawn, then in the triangle BCD, BC, CD, and the angle BCD are given, whence DB is found. Then in the triangle ADB, all the three sides are given; wherefore the angle ADB may be found. Next, in the triangle BED, the sides BE, ED, DB are given; therefore the angle EDB, and its supplement PDB are found. Therefore the three plane angles ADP, ADB, PDB, which contain the solid angle at D, are given; whence the inclination of the planes may be found, and therefore the angle which the plane PAD makes with the plane ADB, that is,

the

the angle PAB, such as it would be measured at A. In the same way the azimuth at B may be found.

d. When these calculations are applied in small triangles, they naturally become much more simple. The process now described, contains a general solution of spheroidal triangles, which have one angle at the pole, whatever be the oblateness of the spheroid, and whatever be the magnitude of the triangles.

69. The Artificial Globe is a delineation of the surface of the earth, and the circles belonging to it, on the surface of a sphere, moveable about an axis; it serves to give a correct notion of the figure and proportion of the parts into which the earth's surface is either naturally or artificially divided, as well as to resolve many of the problems of geography, when great accuracy is not required.

A contrivance of the same kind is applied to the heavens. The uses of the celestial and terrestrial globes are fully explained in most of the treatises on Astronomy and Geography.

70. A Map is a representation of the whole, or of a part of the earth's surface on a plane; and

and though less correct, by being less expensive, and more portable than the preceding, it is of much more general use.

It is impossible to delineate on a plane any figure that can accurately resemble one which is extended in three dimensions. A certain degree of resemblance may, however, be obtained, and, in the construction of maps, this has been sought for in two ways; by the *projection* of the spherical surface on a plane, such as it would be seen to the eye situated in a particular point; or by the *developement*, that is, the spreading out of a spherical on a plane surface.

71. The *Stereographic Projection*, is a representation of the surface of a sphere on the plane of a great circle, such as it would appear to an eye situated in the pole of that circle, or in a point 90° distant from every part of its cirumference.

a. It is usual to suppose the eye placed in the equator, 90° distant from the equinoctial points, so that the plane of projection is the equinoctial colure. If the eye is placed in either of the equinoctial points, the plane of projection is a circle at right angles to the former, called (for a reason that will afterwards appear) the *Solstitial Colure*.

The hemisphere concave to the eye, or on the side of the plane opposite to the eye, is first delineated. The

The eye is then supposed placed in the opposite pole, and the other hemisphere is in like manner represented. It is in this way that the Maps of the World are usually constructed.

b. The stereographic projection has these two very remarkable properties.

1. All the circles of the sphere, both great and small, are represented by circles in this projection.

2. Any two circles cut one another in the projection, at the same angle in which they cut one another on the surface of the sphere. Accordingly, the parallels of latitude in this projection cut the meridians at right angles.

These properties contribute much to the simplicity and beauty of the construction, which, however, has this disadvantage, that the same area on the earth's surface, is represented by a much larger area near the equator, and especially towards the edges of the projection, than at a greater distance. Notwithstanding of this, the stereographic projection is well adapted to *Maps of the World*, or of large portions of the globe.

72. The construction called Flamsteed's Projection, (though it is rather a Developement than a Projection), is very well contrived for the representation

representation of smaller portions of the earth's surface.

a. In this construction, a straight line is drawn for the meridian of the middle of the map, on which are marked off equal distances, to denote degrees of latitude. From a point in this line, as a centre, and with a radius that is to the length of the degree of latitude as the cotangent of the middle latitude to an arch of 1 degree, an arch of a circle is described, to represent the middle parallel of latitude. From the same centre are described other arches, through the different points marked off on the meridian of the middle of the map, which represent the different parallels of latitude. On any one of these parallels, equal distances are set off on each side of the middle point, which are to the assumed degree of latitude, as the cosine of the latitude of that parallel to the radius. The degrees of longitude are thus marked on each parallel, and the curves which pass through the corresponding points in the different parallels are meridians. These are curved more and more on retiring from the middle of the map; but unless the extent is very great, they afford a very good representation of the convex surface.

b. This construction has a very remarkable property, viz. that the quadrilaterals in the map, included between the meridians and parallels of latitude, have the same ratio to one another nearly, with the quadrilaterals which they represent on the surface of the

the globe. See *Mémoire sur la Projection des Cartes Geographiques*, par M. HENRY, 4to, Paris, 1810, chap. 8me, p. 55., &c.

73. The construction which is called MERCATOR's Projection, is chiefly used for nautical charts. In it the meridians are parallel lines; the degrees of longitude are all equal; the parallels of latitude are also parallel lines; and the degrees of latitude increase on the chart in the same ratio that the degrees of longitude diminish in the sphere, or in the spheroid.

a. The consequence of making the degrees of latitude increase, in the manner described, is, that the degrees of latitude and longitude bear to one another the same ratio in this chart, that they actually do on the surface of the earth: and as the meridians are all parallel, the rhumb-lines, or the lines of azimuth, are straight lines. Hence the great use of this construction in navigation.

b. This very ingenious contrivance is alluded to, though obscurely, by PTOLEMY. It was first used among the moderns by MERCATOR, whose name it bears; but the principle of it was first explained by ED. WRIGHT in 1599, who shewed that the parts into which the meridian is divided, must be inversely as the cosines, or directly as the secants of the latitude; and he taught how it might be constructed

structed by the addition of the secants of a series of arches taken in arithmetical progression.

It was afterwards observed, that the meridian line thus divided, was analogous to a scale of logarithmic tangents of the half-complements of the latitudes; this was at first only known as a fact, but was afterwards demonstrated by JAMES GREGORY, in his *Exercitationes Geometricæ*, 1668. Dr HALLEY proved the same in a more concise manner, *Phil. Trans.* N° 219., and greatly improved the construction of the chart.

For other methods of constructing maps and charts, see VARENIUS, Sect. iv. chap. 32. *Encyclopédie Methodique*, art. Cartes. LORGNA, *Principii di Geografia*, 4to, Verona, 1789. *Traité de Topographie*, &c. par L. PUISSANT, Liv. II. LA GRANGE, *Mém. de Berlin*, 1779.

SECT.

SECT. V.

OF PARALLAXES.

HAVING obtained an accurate notion of the figure of the earth, we are enabled to measure the lines either on its surface or in its interior, which must serve as the *bases* from which, by the rules of trigonometry, we are to deduce the distances of objects observed in the heavens. Though the fixed stars are too far off, to have their distances thus ascertained, there may be others, of which the distances admit of being compared with the diameter of the earth. All distances that are not ascertained by the direct application of a measuring line, are determined on the same general principle; that is, from the change in their angular position, which is made by a known change in the position of the observer. This leads us to consider what is called the *Parallax of an Object.*

74. The parallax of any object in the heavens, is the difference of its angular position, as it would

would be seen from the centre of the earth, and as it is seen from a point on the surface.

The parallax of an object, is therefore the same with the angle which the distance between the centre and a given point on the surface subtends at the object.

Though an object to have no parallax, ought, strictly speaking, to be at an infinite distance, yet it will have no sensible parallax, if its distance is very great compared with the diameter of the earth. An angle of one-fourth of a second may be considered as insensible; so that if the radius of the earth subtend an angle, at the distance of any object, less than one-fourth of a second, that object will be seen, from all points of the earth's surface, in the same position.

Now, an arch of 1″ is .000004848 of the radius; and the fourth of a second is therefore

$$.000001212 = \frac{1}{825082};$$

and therefore, if a body is distant from the earth by 825082 of its semidiameters, it can have no sensible parallax.

Though the centre of the earth is a point from which no observations can be made, yet as it is equally related to all the points on the surface, the positions of the heavenly bodies may be most conveni-

ently

ently referred to it. When a star is seen in the zenith of any place, it is seen in the same position as if it were viewed from the centre.

75. The parallax of a body at a given distance from the centre of the earth, is greatest when the body is seen in the horizon. This is called the *Horizontal Parallax*; and the parallax at any given altitude, or the quantity by which the true altitude is diminished, is to the horizontal parallax as the cosine of the altitude to the radius.

If P be the horizontal parallax, p the parallax at the altitude a, $p = P \times \cos a$.

If r be the radius of the earth, supposing it spherical, and d the distance of the body, $\frac{r}{d} = \sin P$.

When P is very small, $P = \frac{r}{d}$, P being expressed, not in degrees, but in parts of the radius; to have it in degrees, $\frac{r}{d}$ must be multiplied by m, the number of degrees in an arch equal to the radius.

If the horizontal parallax is known, the distance d is known; for $d = \frac{r}{\sin P}$.

The

The distances of bodies having different horizontal parallaxes, are therefore inversely as the sines of those parallaxes, or, when the parallaxes are small, inversely as the parallaxes themselves.

76. If two observers, under the same meridian, but at a great distance from one another, observe the zenith distances of the same star, when it passes the meridian on the same day, they can from thence determine the horizontal parallax.

If the amplitude of the arch of the meridian intercepted between the zeniths of the observers, be called a, and if φ and φ' be the zenith distances observed, making one of them negative in respect of the other, if the star be between the two zeniths,

$$\text{then } P = \frac{\varphi - \varphi' - a.}{\sin\varphi - \sin\varphi'}.$$

a. If A and B (fig. 7.) be the places of the observers, Z, Z′ their zeniths, and C the centre of the earth; S the star, of which the parallax is to be found; if SA, SB, SC be drawn,

$$\text{ang. } ZAS = \text{ang. } ACS + \text{ang. } ASC;$$
$$\text{ang. } Z'BS = \text{ang. } BCS + \text{ang. } BSC;$$

subtracting the lower equation from that above it,

$$\text{ang. } ZAS - \text{ang. } Z'BS = \text{ang. } ACB + \text{ang. } ASB;$$
$$\text{or ang. } ZAS - \text{ang. } Z'BS - \text{ang. } ACB = \text{ang. } ASB;$$
$$\text{or } \varphi - \varphi' - a = \text{ang. } ASB.$$

If

If P be the horizontal parallax,

ang. ASC = P × sin ZAS, and
ang. BSC = P × sin Z′BS; whence
ang. ASB = P (sin ZAS — sin Z′BS)
= P (sin φ — sin φ'). But ang. ASB =
$\varphi - \varphi' - a$; therefore P =

$$\frac{\varphi - \varphi' - a}{\sin \varphi - \sin \varphi'}.$$ Hence, also,

$$P = \frac{\varphi - \varphi' - a}{2 \cos \frac{\varphi + \varphi'}{2} \times \cos \frac{\varphi - \varphi'}{2}}.$$

φ' must be supposed negative, if the star is between the zeniths.

b. If the star does not change its declination, this formula will give the horizontal parallax, though the observer be not on the same meridian, because the meridian altitude of such a star, is the same at all places in the same latitude.

c. If there is a change of declination, and if the observers are not under the same meridian, then, by repeated observations of the zenith distance at any of the places, the star's change of declination for a given interval of time, may be found; and so, from the observed altitude at that place, the altitude at any other place, under the same parallel, may be found, if the difference of longitude is known. Hence the altitude is known at the place where the meridian

meridian of the second observer intersects the aprallel of the first, and thus the parallax is computed as above.

This method of finding the parallax, was applied to that of the moon, by LA CAILLE, who observed at the Cape of Good Hope, and LA LANDE, who observed at Berlin, in 1751. LA LANDE, *Astronomie*, tom. II. § 1650.

77. Two observers being supposed, as in the last article, the parallax of a star may be found, by comparing it with a fixed star, (which has no parallax), at the time of its passage over the meridian.

For if S be the object of which the parallax is to be found, as before; T a fixed star, that passes the meridian at the same time, or nearly at the same time with S; the angles TAS, TBS, may be both measured, and they are together equal to the angle ASB; because AT and BT are parallel.

The angle ASB being thus found, the rest of the calculation is as above.

LA CAILLE and WARGENTEN, the former at the Cape, the latter at Stockholm, employed this method for finding the parallax of Mars. LA LANDE, tom. II. § 1652; VINCE, vol. I. § 158.

78. If

78. If these computations are to be made with accuracy, the spheroidal figure must be taken into account, as affecting the observations in two different ways, viz. by making the semidiameters of the earth unequal, under different parallels, and by making the zenith different from the point in which the semidiameter of the earth, if produced would meet the heavens.

If the radius of the equator be called a, as in the last section, and if r and r' be radii of the earth, computed for A and B, and if the horizontal parallax at the equator $= \pi$, at A the horizontal parallax will be $\pi \times \frac{r}{a}$, and at B, $\pi \times \frac{r'}{a}$; and therefore, supposing the zenith distances ϕ and ϕ' to be found as before, but to be corrected for the angles between the vertical and the semidiameter at each station, then $\pi = \frac{a(\phi - \phi' - a)}{r \sin \phi - r \sin \phi'}$.

79. The parallax affects the position of a body, only by depressing it in the direction of a vertical plane, and therefore does not change its place in azimuth, or in the direction of any plane parallel to the horizon; but it changes the place in respect to circles, which cut the vertical obliquely, such as the circles of decli-

nation

nation and of right ascension. These changes are called, the parallaxes in right ascension and declination.

a. In fig. 8. if PZO be the meridian, P the pole, Z the zenith, S the true place of the star, depressed by its parallax to S′, in the direction of the vertical circle ZSS′; let PS be a circle of declination passing through the true place of the star, and PS′ another such circle, passing through the apparent place. If SM be drawn perpendicular to PS′, MS′ is the parallax in declination, and the angle SPM, the parrllax in right ascension. The triangle SS′M may be considered as rectilineal; and if we make the angle at S, or what is called the *angle of position*, $= y$, the parallax in altitude, or SS′, being $= p$, we have $MS' = p \times \cos y =$ parallax in declination; also $SM = p \times \sin y$; and hence the angle $SPM = p \dfrac{\sin y}{\cos \text{Dec.}} =$ the parallax in right ascension.

b. Hence it may be shown, that when the horizontal parallax is given, the parallax in right ascension is proportional to the sine of the hour angle ZPS.

As the parallax in right ascension is nothing when a body is on the meridian, and increases on either side, it may be found by comparing the difference between the times when the body and a star near it, come to the meridian, with the difference between the times when they come to the same hour circle, near the horizon; and thence the horizontal

tal parallax may be concluded. VINCE's *Astronomy*, vol. I. § 160., &c. See also LA CAILLE, *Astronomie*, § 651., &c.; and LA LANDE, Liv. IX. tom. 3. § 1628, &c. From the difference between the altitudes of the body and of the star, as they appear on the meridian, and as they appear near the horizon, the parallax in altitude may be still more directly determined.

SECT. VI.

MOTION OF THE SUN.

80. THE great orb which is the scource of light and heat, and in whose presence all other luminaries disappear, partakes, with the stars, in the diurnal motion; but the time between its passing the meridian one day, and its passing it the next, is greater than a syderial day, and, at a medium, exceeds it by 3 minutes 56 seconds and a half nearly.

The sun, therefore, appears to go eastward among the fixed stars every day, by an arch corresponding to $3^m\ 56\frac{1}{2}^s$, or by $59'\ 8''.3$. This is the mean rate; the real motion is sometimes greater, and sometimes less, than this quantity.

The

The mean interval of time between the sun's passing the meridian one day, and his passing it the next, is called a *mean solar day.*

The solar exceeds the siderial day by 3^m 56^s 33^t of siderial time; or by 3^m 55^s 55^t of solar time; and the lengths are in the ratio of 1.0027399 to 1.

81. The sun does not move eastward in the plane of the equator, but in the plane of a circle cutting it obliquely in the two opposite points, already referred to, of the Vernal and Autumnal Equinoxes, and making, with it, an angle of 23° 27′ 30″ nearly.

a. That the sun's motion is all in one plane, may be shewn by observing his right ascension and declination, every day at noon, and marking it off upon a globe, on which a great circle, representing the equator, has already been described; or, which is better, by remarking, that the sine of the right ascension has always the same ratio to the tangent of the declination. This could not be, unless the plane passing through the sun and the vernal equinox, made a constant angle with the equator, such that the radius had to its tangent a ratio the same with the preceding.

The circle which the sun thus appears to describe in the heavens, is called the *Ecliptic,* and the angle which it makes with the equator, is called the *Obli-*

quity of the Ecliptic. A Year is the time which the sun takes to describe this circle.

82. During the apparent diurnal revolution of the heavens, the plane of the ecliptic does not remain fixed in its place, like the equator, so as to be always vertical to the same point on the earth's surface. The places, however, over which it passes, and through the zeniths of which the sun must pass in his annual course, are all contained within a zone, extending about 23° 28′ on each side of the equator.

a. This tract, which is called the *Torrid Zone,* is bounded by two parallels to the equator, called the *Northern* and *Southern Tropics;* the latitude of each of which is equal to the sun's greatest declination, or to the obliquity of the ecliptic. From them the sun appears to return back, and hence their name.

83. Each pole of the ecliptic is distant from the corresponding pole of the equator, by an arch equal to the obliquity of the ecliptic; and in the diurnal revolution, they describe the circles called the *Polar Circles.*

These circles are also called the *Arctic* and *Antarctic Circles,* and the spaces within them are the *Frigid Zones.*

The

The *Temperate Zones,* are those contained between the Torrid and Frigid Zones.

84. If a straight line be drawn from the centre of the sun to the centre of the earth, a great circle on the earth's surface, perpendicular to this line, will separate the enlightened hemisphere from the dark one; all the places within certain limits, will, by the revolution of the earth on its axis, be made to cross this circle twice in the course of a solar day; at the one of which points, the sun will appear to rise, and at the other, to set; the circle, also, which separates the light from the darkness, will have a vibration corresponding to the sun's annual motion, by which it will incline alternately toward the north and toward the south.

The circle separating Day from Night, or the light from the dark hemisphere of the earth, is called the *Circle of Illumination.*

a. When the sun is in the equator, the circle of illumination passes through both the poles, and every parallel of latitude is one-half in the light and one-half in the dark; so that day and night are equal. This is the season of the Equinox, and the sun's meridian distance from the zenith of every place is then equal to the latitude.

b. As

b. As the sun passes to the north, the circle of illumination inclines also to the north; the sun approaches nearer to the zeniths of all the places in the northern hemisphere, and retires by the same quantity from the zeniths of all the places in the southern. The days lengthen in the former hemisphere, and shorten in the latter.

c. In the Northern Frigid Zone, over a space equal to the sun's declination, the sun does not set; and over an equal space in the Southern, he does not rise.

d. When the sun arrives at the Northern Tropic, he has approached as near as he can to the zenith of every place between the Northern Tropic and the Pole, and has receded the farthest from the zenith of every place in the southern hemisphere. The days are the longest in all the former of these; the shortest in the latter. Within the whole of the arctic circle, the sun does not set on the solstitial day; and within the whole of the antarctic, he does not rise. It is the Summer Solstice in the one hemisphere; the Winter Solstice in the other.

e. The same vicissitudes take place in a contrary direction, as the sun passes to the Southern Tropic.

f. The four points, the two Equinoxes and the two Solstices, divide the year into the four Seasons. The positions of the circle of illumination at these four times are represented, figs. 9, 10, 11.

g. The

g. The length of the day for any latitude, and any declination of the sun, may be calculated, like the continuance of a star above the horizon, § 27.

h. The determination of the length of the year, requires the time of the sun's crossing the equator to be accurately defined.

85. The time of the sun's crossing the equator, is found by observing his declination at mid-day, for several days before and after the equinox ; so that an equation can be found, expressing the relation between the time and the declination sufficiently exact for a few days. If, in this equation, the declination be supposed equal to nothing, the value of the time corresponding will give the moment of the equinox.

The manner of finding an equation between the time and any quantity determined by observations, made at given intervals of time, is called the *Method of Interpolation.* It is much used in astronomy, and will be afterwards explained. See NEWTON's *Principia*, lib. III. lemma 5. LA LANDE, *Astronomie*, tom. III. § 4113. LA CAILLE, *Leçons Elementaires d'Astronomie*, § 135, &c.

Nearly in the same way, the moment is found, when the sun's declination is greatest either on the north or the south side of the equator. The only difference is, that when the equation is obtained between

tween the time and the declination, the latter must be supposed a *maximum*, or its fluxion must be made equal to nothing. Thence the time of the solstice, and also the declination itself, when a *maximum*, may both be found.

86. The length of the year is determined, by comparing together the time of the sun's being in either equinox, or in either tropic, for one year, (found as above), with the time of his being in the same point for another year, distant from the former by a considerable number of years. The interval reckoned in days, and parts of a day, divided by the number of years, gives the true length of the year.

a. It was soon found, that the year was nearly equal to 365 days. But when two equinoxes, at the distance of 60 years were observed, the interval was found to be nearly 21900 days; this, divided by 60, gives $365\frac{1}{4}$; which is nearly the length of the year.

b. This, however, is too great; and more accurate observation has found it to be $365^{d}.242264$, or $365^{d}\ 5^{h}\ 48'\ 51''.6$. Biot, *Astron.* vol. II. § 40. 2d edit.

87. If

87. If, at any place, the sun's greatest altitude in summer, and his least altitude in winter, be determined, half the sum of these altitudes is the complement of the latitude, and half their difference is the obliquity of the ecliptic.

The observations from which these conclusions are deduced, must be corrected for refraction, according to the rules in Sect. II. They ought also to be corrected for the sun's parallax, which, however, cannot be exactly found by any of the methods yet explained.

88. The Ecliptic is divided into twelve equal parts, which are called *Signs*: it also passes through twelve constellations, which are called the *Twelve Signs*; and the zone of the starry heavens, in which they are contained, is named the *Zodiac*.

The names of the twelve Signs, and the characters by which they are usually denoted, are,

Aries,	♈	Libra,	♎
Taurus,	♉	Scorpio,	♏
Gemini,	♊	Sagittarius,	♐
Cancer,	♋	Capricornus,	♑
Leo,	♌	Aquarius,	♒
Virgo,	♍	Pisces,	♓

The

The manner of referring any point to the ecliptic, of which the position is known with respect to the equator, or of finding its longitude and latitude from its right ascension and declination, has been already explained, § 30.

If the point is in the ecliptic; if, for example, it is the sun itself of which the longitude is required, from knowing the right ascension or declination; we have

$$\text{tan. long.} \odot = \frac{\text{tan. Æ} \odot}{\text{cos Obl. Ec.}}, \text{ or sin long.} \odot = \frac{\text{sin Dec.} \odot}{\text{sin Obl. Ec.}},$$ § 80. The character Æ denotes the Right Ascension.

89. The point in which the equator and ecliptic intersect, is not immoveable, but appears, in respect of the fixed stars, to recede towards the west, at the rate of $50''\frac{1}{4}$ nearly *per annum*, or about 1° in 72 years.

This motion is called the *Precession of the Equinoxes*; and by means of it, the equinoctial points describe an entire circle in 25867 years, which is the *annus magnus* of the ancients. In consequence of this, the longitude of the stars continually increases, at the rate of a degree in 72 years nearly. When the zodiac was first delineated by the ancient astronomers, the middle of the constellation Aries was at the vernal equinox, from which it is now distant more

more than 58° towards the east. Hipparchus discovered the precession of the equinoxes, by a comparison of his own with more ancient observations.

90. On account of the precession of the equinoxes, the tropical year, or the time in which the sun moves from the vernal equinox to the vernal equinox again, is less than the time in which he moves from one star to the same star again; the vernal equinox having gone westward so as to meet the sun.

The time in which the sun goes from one fixed star to the same fixed star again, is called a *Siderial Year*, and is longer than the tropical year, by the time that the sun takes to move over $50''\frac{1}{4}$ of his orbit. This amounts to .014119 of a day, or $20'$ $19''.9$; so that the siderial year is 365^d 6^h 9^m $1^s.5$.

91. The Obliquity of the Ecliptic is also subject to change, and appears to have been constantly diminishing from the remotest date of astronomical observation: its present rate of diminution is nearly $50''$ in a century.

A Chinese observation of the sun's altitude at the solstices, as ancient as the year 1100 before Christ, has

has been preserved; and from it La Place deduces the obliquity at that time, = 23° 54′ 2″. *Connaissance des Tems*, 1811, p. 432.

A series of observations, from the age of Pytheas, down to the present time, confirms the same general result. La Place, ibid. La Lande, *Astron.* § 2738.; Vince, *Astron.* I. § 151.

The mean obliquity for 1750, was determined very exactly by La Caille and Bradley, to be 23° 28′ 19″.

The obliquity, beside this progressive diminution, is subject to slight periodic irregularities, which are not here considered. The diminution itself, though apparently progressive, will be found afterwards to be really periodical, and a part of a slow vibration, by which the obliquity of the ecliptic alternately increases and diminishes within very narrow limits.

Apparent

Apparent Orbit of the Sun.

92. If the sun's motion in the ecliptic be determined by observation from day to day, (§ 81. *a.*), it will be found, that it is not uniform, but is swiftest about the beginning of January, and slowest about the beginning of July; being continually retarded from January to July, and accelerated from July to January.

93. If the apparent diameter of the sun be also observed every day by a micrometer, or any instrument that measures small angles with great exactness, it is found to vary, so as to be greatest when the angular motion is greatest, and least, when it is least. The angular velocity, and the diameter of the sun, do not, however, vary in the same ratio; but the angular velocities, at any two points of time, are as the squares of the apparent diameters.

Thus, if v and v' are the angular velocities of the sun, or his diurnal advances in the ecliptic, at any two seasons

seasons of the year, d and d' his apparent diameters at the same times, $v : v' :: d^2 : d'^2$, or $\frac{v}{v'} = \frac{d^2}{d'^2}$.

94. As the apparent diameters of the sun must be inversely as his rectilineal distances from the earth; therefore the apparent velocities of the sun are inversely as the squares of his distances from the earth.

If D and D' be the real distances of the sun from the earth, at the two instants, when the angular velocities are v and v', $v : v' :: D'^2 : D^2$; so that $v D^2$ is a constant quantity; that is, the product of the diurnal motion into the sun's distance from the earth remains always the same.

95. As it is easy to prove that $v D^2$ is double of the sector which the line drawn from the sun to the earth describes, while it moves through the angle v; therefore the areas described by the line drawn from the sun to the earth, in equal increments of time, are equal, and, in any times whatever, are proportional to the times.

This is the First Law of KEPLER, in as far as relates to the motion of the sun.

96. On

96. On comparing the sun's diameter, measured from time to time, with his place observed in the ecliptic, it is found, that if the sun's mean apparent diameter be m, his least diameter $m - n$, and z, his angular distance at any time from the point in the ecliptic where his diameter is least, his apparent diameter at that time is $m - n. \cos z$.

$m = 32' \, 06''.2$, and $m - n = 31' \, 32''.8$, so that $n = 32''.4$, and $m : n :: 19262 : 324$, or as 59.45 to 1.

The sun's apparent semidiameter is therefore always expressed by the formula $(32' \, 6''.2) - 32''.4 \times$

$$\cos z = 32' \, 6''.2 \left(1 - \frac{1}{59.45} \cos z\right).$$

Because the distance of the sun and earth must be inversely as the apparent diameter of the sun, therefore, if that distance be called y, $y = \dfrac{B}{m\left(1 - \frac{n}{m} \cos z\right)}$,

where B is a constant but indeterminate quantity. Now, if a be assumed for the mean distance of the sun, and if $B = m a\left(1 - \frac{n^2}{m^2}\right)$, $y = \dfrac{a\left(1 - \frac{n^2}{m^2}\right)}{1 - \frac{n}{m} \cos z}$;

an

an equation to an ellipsis expressing the relation between y or FG, (fig. 12.) the radius vector drawn to the focus, and the angle z or AFG which that line makes with the transverse axis on the side of the centre.

97. The orbit of the sun is therefore an ellipsis of which the semi-transverse axis is a, the eccentricity $\frac{n}{m}a$, and the semi-conjugate axis $a\sqrt{1-\frac{n^2}{m^2}}$.

Thus the *species* of the sun's orbit is determined. For the property of the ellipsis quoted above, see *Geometry of Curve Lines*, (Prof. LESLIE,) Prop. 22. FRISIUS *de Anal. Sect. Con.* Prob. 29. It may also be easily derived from SIMSON'S *Con. Sect.* Prop. 4. Book 2.

98. The position of the transverse axis, or of what is called in astronomy, the *Line of the Apsides*, in respect of the fixed stars, is found, by taking the middle point between two positions of the sun, in which the diurnal motion is the same.

The

The place of the apsides thus found, may be corrected, on the principle, that the axis is the only line drawn through the focus of an ellipse, which bisects its area. The time, therefore, from the sun's being in one apsis, to his being 180° farther advanced, is precisely half a year. When, therefore, two points are found in the sun's orbit, 180° or 6 signs distant from one another, and such, that the sun takes precisely half a year to pass from the one to the other; they are the Higher and the Lower Apsis. LA CAILLE, *Mém. de l'Acad. des Sciences*, 1742. Also his *Astronomie*, p. 78. WOODHOUSE, *Astron.* p. 210.

Thus the longitude of the aphelion, for the year 1780, was determined by DE LAMBRE, from Dr MASKELYNE's observations, to be 99° 8′ 19″.9, or 9° 8′ 19″.9 advanced into the sign Cancer. The sun passed through this point 4m past noon on the 31st of June, according to the time at Greenwich. BIOT, *Astron.* vol. I. p. 220.

The eccentricity of the sun's orbit was found, in (§ 97.) $= \frac{n}{m} \times a$. A method of finding $\frac{n}{m}$ more accurately than can be done merely by the observations of the sun's diameter will be immediately explained. When so determined, it is found $= .016814$, the mean distance being 1. In the following propositions, the mean distance $= a$, and the eccentricity $= e$.

99. The

99. The position and the species of the orbit, being thus determined, the calculation of the sun's place for any given time, is reduced to the geometric problem, of drawing a line through the focus of an ellipsis, so as to cut off a sector between it and the transverse axis, having a given ratio to the whole elliptic area.

If AGPH (fig. 13.) be the orbit of the sun; F the focus in which the earth is placed; and G the place of the sun at a given time; then the time of the sun's describing the arch PG, or of the radius vector describing the sector PFG, is given; and the time of an entire revolution, or of the radius vector describing the whole elliptic area, being also given, the ratio of the sector PFG, to the whole ellipsis AGPH, is given. If, from this, the position of the straight line FG can be determined, it is evident, that the angle PFG, and the position of G, are found.

This is known by the name of KEPLER'S *Problem*; it can only be resolved by approximation.

a. The angle PFG, which measures the angular distance of the sun from the *Perigee*, or lower apsis, is called the *true anomaly*.

b. If a circle be described from the centre F, with a radius $= \sqrt{a\,(a^2 - c^2)^{\frac{1}{2}}}$, its radius will be equal to

1 that

that of the ellipsis; and if a sector *a* FG′ be taken in the circle, equal to AFG in the ellipsis, the angle AFG′ is called the *mean anomaly*.

This angle is proportional to the time, and increases at the rate of 59′ 8″.3 daily.

c. The angle GFG′, or the difference between the mean and the true anomaly, is called the *Equation of the Sun's Centre*.

d. If the true sun G, and the imaginary sun G′, set out together from A and *a*, the sectors AFG and *a* FG′, increasing at the same rate, the angles AFG, and *a* FG′, must increase unequally, and the point G′ must get before G, or the mean anomaly must exceed the true. But when FG becomes less than FG′, the angle AFG will come to increase faster than *a* FG′. At the Perigee they will coincide; and at a certain point, between A and P, their difference, or the equation of the centre, will be the greatest possible. This will happen, when the increments of the true and of the mean anomaly are equal to one another, or when the sun's real angular velocity is equal to his mean, or to 59′ 8″.3.

Among the great number of solutions which have been given of this celebrated problem, one distinguished for the simplicity of the result, proceeds on the supposition, that the angles at the superior focus are proportional to the time, or are the same with the mean anomaly; which is not far from the truth, in elliptical orbits of small eccentricity. This solution was first proposed by BULLIALDUS, a French astronomer, and was afterwards adopted and improved by Dr SETH WARD, and is known in this country by the name of WARD's *Hypothesis*. Another solution is distinguished, for the simplicity of the principles, and the elementary nature of the reasoning employed, viz. that given by the late Dr MATHEW STEWART, in the *Edinburgh Physical and Literary Essays*, vol. II. (1755) p. 105.; and again republished in his *Physical and Mathematical Tracts*, p. 404. Among the other solutions, those of NEWTON, *Prin. Math.* lib. I. prop. 30. Schol.; of SIMPSON, *Essays*, 4to, (1740) p. 47.; of EULER, *Comment. Petrop.* tom. VII.; and of IVORY, *Edinburgh Transactions*, vol. V., (the latter extending to the most difficult case of the problem, when the eccentricity is great), are particularly commendable.

Of all these, however, it may be said, that though excellent when a numerical calculation only is required, yet when the solution is to be a step in the investigation of other properties of the elliptic motion, they cease to be of use, so that recourse must be had to such general theorems, as express the true anomaly in terms of the eccentricity, and of the mean anomaly. The first solution of this kind was given by CLAIRAULT, *Theorie de la Lune*, § 31. lemma 3d, &c. It was afterwards improved and extended by other mathematicians, particularly ly BOSSUT and JEAURAT. See also LA LANDE, *Astronomie*, tom. III. § 3481.; and CAGNOLI, *Trigonometrie*, § 1488. The series was continued by JEAURAT, as far as the 9th power of the eccentricity.

100. It is proved in that solution of KEPLER's Problem, that if 1 be the semitransverse, and e the eccentricity of an elliptic orbit, x the mean anomaly, reckoned from the Perigee, and y the equation of the centre;

$$y = +\left(2e - \frac{1}{4}e^3 + \frac{5}{96}e^5\right)\sin x$$

$$+\left(\frac{5}{4}e^2 - \frac{11}{24}e^4 + \frac{11}{192}e^6\right)\sin 2x$$

$$+\left(\frac{13}{12}e^3 - \frac{43}{64}e^5\right)\sin 3x$$

$$+\left(\frac{103}{96}e^4 - \frac{451}{480}e^6\right)\sin 4x$$

$$+\frac{1097}{960}e^5\sin 5x + \frac{1223}{960}e^6\sin 6x.$$

This value of y, applied to the mean anomaly according to its sign, will give the true anomaly.

a. When the equation of the centre is found from this formula, the constant coefficients must be reduced into degrees and minutes, by multiplying each of them

them by 57°.29578, &c. the number of degrees in an arch equal to the radius.

b. Because, in the case of the sun, e is small, (viz. $= .016814$), its powers above the third may be rejected; and so the equation reduced is,

$$y = (1^\circ\ 55'\ 26''.35) \sin x + (1'\ 12''.68) \sin 2x.$$
$$+ (1''.05) \sin 3x.$$

101. The *Radius Vector* r, or the line drawn from the sun to the earth, may also be expressed in terms of the mean anomaly, supposing the mean distance = 1.

$$r = 1 + \frac{1}{2}e^2 - (e - \frac{3}{8}e^3 + \frac{5}{192}e^5) \cos x$$
$$+ (-\frac{1}{2}e^2 + \frac{1}{3}e^4 - \frac{1}{16}e^6) \cos 2x$$
$$-\left(\frac{3}{8}e^3 - \frac{45}{128}e^5\right) \cos 3x$$
$$+ (-\frac{1}{3}e^4 + \frac{2}{5}e^6) \cos 4x$$
$$-\left(\frac{125}{384}e^5\right) \cos 5x$$
$$+ (-\frac{27}{80}e^6) \cos 6x,$$

The

The series is here, and in the preceding case, adapted to the computation of x from the perigee. When x is computed from the apogee, the signs of the terms involving the *odd* multiples of x, must be changed. A small part of this series will give the radius vector in the case of the sun, with sufficient accuracy; we may suppose

$$r = 1 + \frac{1}{2}e^2 - e \cos x - \frac{1}{2}e^2 \cos 2x.$$

102. Though the eccentricity is supposed to be found, from § 97., it must be corrected by means of the greatest equation of the centre, determined by observation; this equation being called g, if

$$\frac{g}{57.29578} = h; \text{ then } e = \frac{1}{2}h - \frac{11}{768}h^2 - \frac{587}{983040}h^5 - \&c.$$

As h is very small in the sun's orbit, being $= .033629$, $e = \frac{1}{2}h$ nearly; that is, $e = .016814$.

Syst. du Monde, p. 116.; Biot, *Astron. Phys.* tom. II. p. 228. 2d edit.

a. The greatest equation itself is found by observing the sun's place in the ecliptic, day after day, and comparing it with the mean place, calculated as in § 99. *b*.

103. Astronomical

103. Astronomical tables, constructed from the data, and on the principles now explained, serve to determine the sun's place in the ecliptic, for any instant of time, either past or future.

a. From the time of the sun's passing through the perigee, when the true and mean place coincide, the mean place for any other time may be computed, by allowing for the interval an increase of longitude, at the rate of 59′ 8″.3 *per diem*, and thus the mean anomaly is computed. From the mean anomaly, is found the equation of the centre, contained in a table, which gives the quantity of that equation for every degree of mean anomaly; thence is given the true anomaly, and of course the true longitude. The tables are so constructed, as to give the mean place of the sun for the beginning of every year; hence the mean place for any time of any year is easily found. When *equations* are thus ranged in tables, the quantity by which they are found out in the table, and on which their magnitude depends, are called the *Arguments of the Equations*.

The tables of the sun's motion, are in reality nothing else than the expansion of the general formula contained in § 100.

b. Thus, from the fact of the variation of the sun's apparent diameter, compared with the variation of the angular velocity of the same body, it is demonstrated, stratad, that, whether it be the earth or the sun which really moves, the one describes an ellipsis round the other, placed in the focus of that curve; and that the line drawn from the moveable to the immoveable body, (the radius vector), describes areas round the latter proportional to the times. From these two general laws, astronomers have been able to determine the proportional distances of the sun and earth, and their relative positions, at all seasons of the year.

c. The use of facts, in scientific investigation, could not be better illustrated than by this example. The simple fact, that the sun on a particular day had a certain apparent diameter, and a certain rate of angular motion, is nothing in itself, and cannot, taken alone, lead to any conclusion: But a multitude of such facts, compared together, by the assistance of geometry, leads to the knowledge of the general law to which they are all subject; this law leads, conversely, to a more precise and comprehensive knowledge of facts, embracing at once the past, the present and the future, and reducing them all into one theorem.

We are left, however, in the dark as to one question, whether it be the sun or the earth which describes the elliptic orbit. The appearance that the earth is at rest, and that the sun is in motion, is no argument against the fact being entirely the reverse, as has already been made evident in the case of the

the earth's rotation. The decisive facts have not yet occurred, which are to determine, whether or not a motion of translation belongs also to that body.

In order that the theory of the elliptical motion may have its conclusions extended either to the future or the past, a very accurate measure and reckoning of time are necessary, and we are now in possession of the *data* by which these may with certainty be obtained.

Of the Equation of time, and of the Kalendar.

The arrival of the Sun in the meridian, being a more conspicuous phenomenon than that of a Star, has been taken to mark the beginning and end of the day, used for the purposes of civil life. Solar time, consists of days measured in this manner, and is used by astronomers, as well as by the people at large. Astronomers begin the day at noon, and reckon 24 hours round to noon again : in the common reckoning, the day begins at midnight, and is divided into 24 hours, which are counted by 12 and 12. La Place has proposed, that in this the astronomers should follow the people, as, by beginning the day at midnight, the whole of the sun's stay above the horizon falls in the same day.

104. The

104. The difference between the solar and siderial day, has already been shewn to consist of 3^{m} 56^{s} 33^{t} of siderial, or 3^{m} 55^{s} 49^{t} of solar time; but the solar day thus defined, is only the average, or the mean; for, the time actually employed by the sun to return from the meridian to the meridian again, is different at different seasons of the year, § 92.

105. If the sun moved regularly forward in the equator, at the rate of 59′ 8″.3, every day, the solar days would be all equal, and their lengths as above determined; but as the sun neither moves in the equator, nor moves in the ecliptic, at a uniform rate, there are two causes that affect the length of the solar day.

The time, therefore, which is reckoned by the arrival at the meridian of an *imaginary sun*, or one which is supposed to move in the equator, is called *mean solar* time. When it is reckoned by the arrival the real sun on the meridian, it is called *apparent* time.

106. The difference between the right ascension of the sun, and his mean longitude, converted into mean solar time, is the difference between

tween the mean and the apparent time, and is called the *Equation of Time*.

a. The conversion of degrees and minutes, into mean solar time, is performed by a different rule from their conversion into siderial time. For an hour of mean solar time, there must be reckoned 15° 2′ 27″.847 of the equator; and so on, in the same proportion.

b. There are four times in the year, when the mean longitude of the sun, and his true right ascension, are equal to one another; and, at these times, the apparent and the mean time coincide. These times happen, at present, about the 15th of April, the 15th of June, the 1st of September, and the 24th of December. From the first of the above periods to the second, the apparent time is before the mean, and the equation of time is subtractive, or must be taken from the apparent time, to give the mean. It is greatest at the 15th of May, when it amounts to 3′ 58″ to be subtracted. From the second to the third interval, the equation is additive, the mean time being before the apparent; and it becomes a maximum about the 25th of July, when it amounts to 6′ 6″. It becomes negative between the third and fourth interval, and reaches its *maximum* on the 2d or 3d of November, when it amounts to 16′ 15″ subtractive. Dr Halley has given a geometrical construction, for determining the time when the apparent days are longest or shortest, and also when they are equal to the mean.

mean. See KEIL's *Astronomy*, Sect. 25. p. 322., &c.

c. When the equation of time is a *maximum*, the length of the mean and apparent day is the same. When the equation of time is nothing, the lengths of the apparent and the mean day differ most from one another. On the 16th of June, for example, when the equation of time vanishes, the day is 12″.5 shorter than the mean solar day. The difference becomes less and less till the 26th or 27th of July, when the equation of time is stationary or a *maximum*, being then $6^m\ 5^s$ nearly; and the apparent day is then in length equal to the mean. The difference of the days increases from thence; and on the beginning of September, the apparent day is 18^s longer than the mean, &c.

d. Clocks ought to be regulated by the mean solar time; and when they are adjusted by the sun's passing the meridian, the equation of time must be applied.

e. The mean length of the day is thus accurately determined. But in the reckoning of time for the purposes of astronomy, and of civil life, we must not count by days only, but by that assemblage of days which constitutes a year, and which is naturally pointed out as a division of time by the return of the seasons. The sort of incommensurability that exists between the lengths of the day and of the sun's revolution, renders it somewhat difficult

difficult to adjust the reckoning of both, when it is to be done in whole numbers.

107. If the revolution of the sun consisted of an entire number of days, for instance 365, the year would naturally be made to do the same, and there would be no difficulty in the formation of the *Kalendar*, or in adjusting the reckoning in years and in days to one another.

All the years would then be precisely of the same number of days, and would all begin and end with the sun in the same point of the ecliptic. If the time of the sun's revolution includes a fraction of a day, the case is different; a year, and a revolution of the sun, cannot then be precisely of the same duration.

108. As it would be inconvenient to begin the year at any point of time, but the beginning of a day, a year must consist of an entire number of days. If, therefore, the revolution of the sun include a fraction of a day, for example one-fourth, then one year cannot be made equal to one revolution of the sun; but four years may be made equal to four revolutions of the sun.

Suppose

Suppose, for example, that $365d + \frac{1}{4} R =$, a revolution of the sun; then

$4 \times 365 + 1 = 4 R = 3 \times 365 + 366$; so that if we count three years, each of 365 days, and a fourth of 366, we shall have exactly four revolutions of the sun; and at the beginning of the fifth year, the sun will be in the same point of the ecliptic which he was in at the beginning of the first. This is the arrangement of what is called the *Julian* Kalendar; and if the revolution of the sun were exactly $365^d\ 6^h$, it would be altogether perfect. It is called the *Julian* Kalendar, and the Year thus computed, the *Julian* Year, from JULIUS CÆSAR, by whom, with the advice of the astronomer SOSIGENES, it was introduced at Rome.

The addition of a day, or a number of days, to any fixed period, at stated intervals, is called *Intercalation.* The year on which the intercalation fell was called *Bissextilis,* because the 6th of the *Kalends* of March was twice counted. With us, it is called *Leap* year.

109. As the true length of the sun's revolution is not what has now been supposed, but instead of $365^d.25$, is only $365^d.242264$, the Julian year is longer than the revolution of the sun by $0^d.007736$, (nearly 11^m); and, therefore, before a new year begins, the sun has passed the point in

in the ecliptic where the last year began, by a small fraction, viz. .007736 × 59′ 8″.

The Julian reckoning, therefore, falls continually behind the sun, and the course of the seasons, by a quantity which, however, is so small, that it was long before it was observed.

a. At the time of the *Council of Nice*, in the year 325 of the Christian era, the Julian Kalendar was introduced into the Church; and at that time the vernal equinox fell on the 21st of March. On account of the imperfection in the mode of reckoning just noticed, the reckoning fell constantly behind the true time; so that in the year 1582, the Julian year had fallen nearly 10 days, (9.72415) behind the sun; and the equinox, instead of falling on the 21st, fell on the 11th of March; so that the difference was nearly a day in 132 years. The continuance of this erroneous reckoning, would have made the seasons change their places altogether; and it was therefore resolved to reform the kalendar, which was done by Pope Gregory XIII., and the first step was to correct the loss of the ten days, by counting the day after the 4th of October 1582, not the 5th, but the 15th of the month.

110. As the loss in the Julian Kalendar amounted to one day in 132 years, it would amount to three, in 396 years, or in the space nearly of four centuries. It would be necessary,

sary, therefore, supposing the Julian intercalation to continue, to suppress three intercalary days in the course of four centuries; and it was agreed, that this should be done on the three successive secular years, retaining the intercalary day on the fourth, by which means the sun, at the beginning of the fifth century, would occupy the same point in the ecliptic, within a few minutes, that he did at the beginning of the first.

a. The amount of the accuracy thus attained may be more exactly estimated. The annual error amounts to .007736, which, in one century, is .7736 of a day, and in four centuries, 3d.0944. Of this error, the fractional part 0d.0944, or nearly 2h 15m, is all that remains uncorrected by the GREGORIAN reformation, and in this time the sun describes an arch of 5′.½ in the ecliptic. The sun, at the beginning of the fifth century, will therefore be only 5′.5 advanced beyond the point he was in at the beginning of the first.

b. The reformation of the kalender did not take place in England till the year 1752. The secular years on which the intercalary days are omitted, are 1700, 1800, and 1900.

c. Though this degree of accuracy is sufficient for the purposes of history, and even of astronomy, and may easily be carried farther, by similar means adopted in future ages, yet the problem of intercalating, so that the difference between the computations

tions in the kalender, and the real motions of the sun, should always be the least possible, is not thereby completely resolved. The modes of intercalation best suited to that object, require all the integer numbers to be found which most nearly express the ratio of the fraction .242264 to 1. See EULER, *Elemens d'Algebre*, tom. II.; Additions by LA GRANGE, § 20.

Secular Variations in the apparent Motion of the Sun.

The variations in the sun's motion which have now been described, are confined within short periods, during which they alternately increase and diminish. There are others, which go on from one age to another. and are either continually progressive, or circumscribed by periods of very long duration. These are so slow, that they are only perceived by comparing together observations made at a great distance of time. They are called *Secular Inequalities.*

111. By comparing very distant observations, it is found that the line of the *apsides*, or the longer axis of the sun's orbit, has a progressive motion, or a motion eastward: so that the apsis recedes

recedes from the vernal equinox 62″, or by De Lambre's Tables 61″.9 annually.

a. This motion includes the precession of the equinoctial points, which is in the opposite direction, and amounts to 50″.25; so that the real motion of the apsides eastward, in respect of the fixed stars, is 11″.65 a-year, or 19′ 4″.$\frac{1}{6}$ in a century.

b. Hence there is a difference between the Tropical Year, or the time of the sun's revolution from equinox to equinox, and what is called the *Anomalistic* Year, or the time of the sun's revolution from either apsis to the same apsis again. As the apsis has gone in the same direction with the sun over 62″ in a year, the sun must come to the place where the apsis was at the beginning of the year, and must move over 62″ more before the anomalistic year is completed. The time required to this is .01748 of a day, which, added to the tropical year, gives $365^d.259744$, or $365^d\ 6^h\ 14^m\ 2^s$ for the anomalistic. Biot, *Astron.* tom. II. § 91.

c. The line of the apsides thus continually moving round, must at one period have coincided with the line of the equinoxes. The lower apsis or perigee in 1750, was 278°.6211 from the vernal equinox, according to La Caille; and the higher apsis was therefore at the distance of 98°.6211. The time required to move over this arch, at the rate of 62″ annually, is about 5722 years, which goes back nearly 4000 before our era,—a period remarkable

 for

for being that to which chronologists refer the creation of the world. At that period, the length of time during which the sun was in the northern signs, that is, on the north side of the equator, was precisely the same with that on which he was on the south, each being exactly half a year. At present, the apogee, where the sun's motion is slowest, being in the ninth degree of Cancer, more time by $7^d\ 16^h\ 30^m\ 8^s$ is consumed in the northern than the southern signs; so great is the change which the motion of the apsides has produced. About 464 years ago, the apogee was in the beginning of Cancer.

d. When we would calculate the place of the sun for any time, (especially if very distant), we must begin with determining the position of the line of the apsides; as it is from the lower apsis that the mean and true anomaly are both computed.

e. The motion of the sun's apsides being $19' 4''$ in a century, with respect to the fixed stars, it requires a period of more than 108000 years to complete their siderial revolution. Their tropical revolution is 20903 years.

112. By a comparison of distant observations, the equation of the centre, and of consequence the eccentricity of the sun's orbit, are discovered to be continually diminishing.

The

The rate of diminution in the *maximum* of the equation, is about 18″.79 in a century.

If the rate of this diminution were to continue uniform, (which, however, we have not a right to suppose), the earth's orbit would become a circle in about 36300 years.

113. From the changes in the place of the apsides, in the equation of the centre of the sun's orbit, and in the obliquity of the ecliptic, the equation of time undergoes a change from one age to another.

a. The equation of time is equal to the mean longitude, *minus* the true right ascension of the sun. The former of these is not affected by the place of the apogee or perigee, and depends only on the distance of time from the equinox. The latter, the true right ascension, is dependent on all the three elements just mentioned. The true longitude of the sun, from which the right ascension is deduced, depends on the place of the perigee, and on the equation to the centre, both of which vary; and it involves also the reduction from the ecliptic to the equator, which depends on the angle these circles make with one another. The equation of time, therefore, answering to the same day of the year, varies in different ages; but for any age, past or future, may be computed on the principles already explained.

b. The

b. The secular variation of the equation of time, is different for every different state of that equation: it is greatest about the time when the sun is in the perigee; and it is then $15''.6$.

The Sun's Rotation on his Axis.

114. The face of the sun, when viewed with a telescope, though of a bright and intense light, far above that of any other object, is often marked with dark *Spots*, which, when examined from day to day, are found to traverse the whole surface from east to west, in the space nearly of fourteen days.

a. These spots, though only visible with the telescope, are sometimes so large, as to subtend an angle nearly of one minute. Their number, position, and magnitude, are extremely variable. Each of them is usually surrounded with a *Penumbra*, beyond which is a border of light, more brilliant than the rest of the sun's disk. When a spot is first discovered on the eastern limb, it appears like a fine line; its breadth augments, as it approaches the middle of the disk, from which it diminishes as it goes over to the western limb, where, at last, it entirely disappears. The same spot, after fourteen days, is sometimes discovered again on the east

east side. It is not often, however, that this happens, as the spots dissolve and perish, from causes that are to us quite unknown.

b. The spots of the sun were first discovered by GALILEO at Florence in 1611. *Istoria e Demonstrazione intorno alle Macchie Solari. Opere di* GALILEO, tom. II. p. 85., &c. *Edit. di Padova*, 1744.

115. The position of a spot, relatively to the sun's centre, may be determined by observing the difference between the time when the centre of the sun passes the vertical wire of a transit telescope, and when the spot does the same. This gives the difference in right ascension; and the difference in declination may be determined at the same time, by means of the moveable wire in the telescope. In this way, by observations made day after day, the path of the spot on the sun's disk may be traced with great exactness.

a. When the spots are followed in this way, many of them are found to change, and to disappear altogether in the course of a few days. Sometimes a number of small spots unite into one large spot; at other times, a large spot separates in a number of small ones, which soon disappear entirely. A few of them, more permanent than the rest, may be followed by the observation here described, as long as two or three revolutions.

116. The

116. The Paths of the spots thus traced, are observed to be rectilineal at two opposite seasons of the year, the beginning of June and the beginning of December, and to cut the ecliptic nearly at an angle of 7° 20′. Between the first and the second of these seasons, the paths of the spots are convex upward, or to the north, and acquire their greatest curvature about the middle of that period. In the following six months, the paths of the spots are convex toward the south, and go through the same series of changes. They appear to be elliptic arches.

In the beginning of March and September, when the opening of the elliptic paths is at its *maximum*, the smaller axis is to the greater as 13 to 100. LA LANDE, *Astronomie*, 3233.

117. The preceding appearances may be explained, by supposing the spots to be opaque bodies, attached to the luminous surface of the sun; the sun having a revolution on an axis inclined at an angle of 7° 20′ to the axis of the ecliptic.

The apparent revolution of a spot is performed in 27 days; but in this time the spot has done more than complete an entire revolution, having, in addition to it, gone over an arch equal to that which the sun has described in the same time in his orbit. This reduces

reduces the time of the sun's rotation on his axis to $25^{d}\ 9^{h}\ 36^{m}$. All the observations do not agree exactly in bringing out this conclusion. LA LANDE, § 3276., &c.

The problem, From three positions of a spot on the sun's disk, in respect of the ecliptic, to find the position of his equator, and of his axis of rotation, is of considerable difficulty. It was first directly resolved by BOSCOVICH in 1737. See LA LANDE, § 3257, who has also given a solution. Dr HERSCHEL has made many important observations on the solar spots. *Phil. Trans.* 1794, 1801.

118. A phenomenon, from which no information concerning the motion of the sun has yet been obtained, though, from some circumstances, it appears to be connected with his rotation, is the *Zodiacal Light;* a luminous appearance, seen after sun-set, or before sun-rise, somewhat similar to the milky way, but of a fainter light, in the figure of an inverted cone or pyramid, with its base toward the sun; having its axis inclined to the horizon, and nearly in the plane of the ecliptic.

The Zodiacal Light was discovered by CASSINI in 1683; but there is reason to think, that it had been remarked before that period.

The

The length of the Zodiacal Light, taken from the sun upwards to its vertex, is various, from 45° to 100°, and even 120°.

The season most favourable for observing this phenomenon, is about the beginning of March, after sun-set: the axis extends toward Aldebaran, and makes, with the horizon, an angle nearly of 64 degrees.

The aspect of the Zodiacal Light is by no means uniform; it is much brighter in some years than others. A remarkably brilliant appearance of it was observed at Paris, 16th February 1769.

The Zodiacal Light appears to be inclined to the ecliptic at an angle of 7°, and to cut it in the 18th of Gemini; so that it is in the plane of the sun's equator, or perpendicular to his axis of rotation. LA LANDE, *Astron.* § 847.

As the decisive facts have not yet occurred, which are to determine whether or not the apparent motion of the sun is to be explained by the real motion of the earth, the language hitherto employed is entirely that suggested by the appearances. The lower apsis of the elliptic orbit, or that nearest to the focus in which either

ther the sun or the earth is placed, has been called the *Perigee*; supposing that it is the point where the sun in motion approaches nearest to the earth at rest. On the contrary supposition, it would be called the *Perihelion*. The same applies to the terms *Apogee* and *Aphelion*.

It must be observed, that the distances of the Sun and Earth, now treated of, are only the relative distances, referred to an imaginary standard, and not compared with any known magnitude, such as the diameter of the earth. The observations on which such a comparison as this must be founded, and by which the sun's parallax is ascertained, cannot be explained till the planetary system is understood. In the mean time, it may be remarked, that even by help of the observations above described, it may be inferred, that the sun's parallax is less than 10″.

From observations of the sun's solstitial altitudes, we saw, § 87., how the latitude may be deduced. In order to make the deduction with accuracy, the sun's altitudes must be corrected for the parallax; and it may be easily shewn, that if the horizontal parallax of the sun is so great as 10″, the lat. of Greenwich, for example, determined without making the due allowance for it, would be erroneous, by 7″. The latitude determined by observations of the stars, is free from this error, and therefore ought not to agree with the former by 7″. Now, the correction necessary to be made, in order that the determination deduced from the altitude of the

sun,

sun, should agree with that from altitudes of the stars, is not easily fixed with perfect precision, but is certainly not so great as 7″. The sun's horizontal parallax is therefore less than 10″.

If the correction above mentioned could be found with perfect precision, this method of determining the sun's parallax, on account of its simplicity, would be preferable to all others. Though it does not afford a precise measure, it assigns a limit that may be used till a more exact result can be obtained.

The parallax of the sun being less than 10″, his distance must exceed 20626 semidiameters of the earth; and his diameter must be to that of the earth in a greater ratio than that of 15′.5 to 10″, or of 93 to 1. § 75.

SECT.

SECT. VII.

MOTION OF THE MOON.

119. THE Moon, next to the sun, is the most remarkable of all the heavenly bodies, and is particularly distinguished by the periodical changes to which its figure and light are subject. These changes are called the PHASES of the Moon.

Modò curvata in cornua, modò æqua portione divisa, modò sinuata in orbem, maculosa, eademque subitò prænitens, immensa orbe pleno, et repentè nulla, &c. PLIN. *Hist. Nat.* lib. II. cap. 9.

120. At the time when the moon is due south, about midnight, her disk is an entire circle, subtending an angle not much different from half a degree. On the next night, she comes later to the meridian by about 48 minutes, and the western part of her disk is no longer bounded by a circle, but by an elliptic line; and this line, every subsequent night, is seen to encroach more on the luminous part, till, on the seventh night,

night, it is nearly a straight line, and the disk a semicircle. The diminution continues, the disk becoming more and more concave to the west, till, about the end of another seven days, it disappears altogether.

After a few days, the moon again appears, like a fine crescent, to the eastward of the sun, with its concavity turned toward the east, and continues to increase on that side till it become entirely full orbed, about 29 days from the time when it was last full.

The line separating the light and the dark part of the moon is irregular and serrated, and its form varies while one is looking at it through the telescope: the light, as it advances, touches some points, while they are yet at a distance from the illuminated surface, and while all round them is dark. The light on them spreads, till it be united to the rest.

The Moon, when full, is opposite to the Sun: when she disappears, or when it is New Moon, she is in conjunction with the sun: these two aspects of the moon are called the *Syzygies.* At the time when the moon appears as a semicircle, she is 90° distant from the sun on either side: she is then said to be in the *quadratures.*

The moon, during all these changes, advances among the fixed stars, at the rate of 13° 10'.½ at an average

rage in 24 hours, and comes later to the meridian by about 48 minutes every day. The motion of the moon is therefore in the same direction with that of the sun, or according to the order of the signs.

121. These Phases may all be explained on the supposition, that the moon is an opaque body, spherical, or nearly so, which moves in a circular orbit round the earth, while it receives its light from the sun.

The opacity of the moon is proved not only from her phases, but from the occultation of stars by the dark or invisible part of her orb.

122. When the place of the moon is observed every night, it is found that her motion is in the plane of a great circle, inclined at an angle (subject to a small variation) of nearly 5° 9′ to the ecliptic.

The line in which the plane of the moon's orbit cuts the ecliptic, is called the *Line of the Nodes.*

The *Ascending Node*, is the point where the moon is in the ecliptic, *ascending* to the north: the *Descending Node*, is that where she is in the ecliptic, *descending* to the south.

123. The

123. The position of the nodes is found, by observing the longitude of the moon, when she has no latitude; and it appears, by a comparison of such observations, that the line of the nodes is not fixed, but has a slow retrogade motion, at the rate of 3′ 10″.64 in a day; so as to complete a revolution relatively to the fixed stars in 6793.4212 days.

Biot, *Astron. Phys.* vol. II. p. 351. 2d edit.

As the moment of the moon being in the ecliptic may not be actually observed, yet, from several observations, taken before and after, the exact time, by the method of interpolations, may be found.

124. By comparing places of the moon, observed at very remote periods, it is found that the secular mean motion of the moon, relatively to the fixed stars, is 1336 circumferences *plus* 306°.48763, which gives for the time of a sidereal revolution $27^{d}.321661$.

In the same way, the tropical revolution, (taking in the precession of the equinoxes), is found to be $27^{d}.321582$; and the diurnal motion relatively to the tropics 13°.18636.

It also has been found, on making these comparisons for very remote ages, that the secular mean motion

was

was less formerly than it is at present. This increase of the mean motion is called the *Moon's Acceleration.* It amounts to 9″ in the present age; and the effect of it on the moon's place increases as the squares of the time.

125. The period of the phases, or the time from full moon to full moon, is greater than the tropical revolution, as being the time required to describe 360°, with an angular velocity that is equal to the difference between the angular velocity of the moon, and the angular velocity of the sun: this gives the period of the phases, or what is called the Lunar Month, equal to

$$29^d.530588 = 29^d\ 12^h\ 44^m\ 2^s.8.$$

See Biot, tom. II. p. 358. Also Woodhouse, p. 305. The above is La Place's determination.

Moon's Orbit.

126. It may be found, from observations of the Moon's apparent diameter, as was done in the case of the Sun, that the orbit in which she moves is nearly an ellipsis, having the earth in its focus, about which the radius vector drawn

to

to the moon describes areas proportional to the time.

a. The same may be also proved, from observations of the Moon's parallax; the Moon being near enough to the earth to admit of her parallax being accurately determined.

b. The Moon's parallax, when least, is $53' \; 50''.99$, and when greatest, $1^\circ \; 1' \; 22''.99$. Hence the Moon's greatest distance is 63.8419 semidiameters of the earth, and her least 55.9164. The mean distance of the moon from the earth is therefore 59.8791, which is also the semi-transverse axis of her orbit.

c. A more accurate determination of these elements, has fixed the mean equatorial parallax at $57' \, 11''.4$; the greatest distance of the moon at 1.05518, the least at .94482, the mean distance being 1.

Hence the eccentricity is .05518.

In round numbers, the moon's mean distance is 60 semidiameters of the earth; and her diameter is to the diameter of the earth as $15'$ to $57' \; 11''.4$, or as 3 to 11 nearly. VINCE, *Astron.* vol. I. § 171.

127. From the eccentricity of the moon's orbit, the equation of the centre is found,

$= (6^\circ\ 18'\ 17''.3) \sin x + (12'\ 49''.7) \sin 2x + (42''.3) \sin 3x$, x being put for the mean anomaly, reckoned from the perigee.

This is the equation of the centre, as deduced from De Lambre's Tables, published in Vince's *Astron.* vol. III. see p. 133., &c.

The *maximum* of the equation is $6^\circ\ 17'\ 28''$, and takes place when the mean anomaly $= 86^\circ\ 5'\ 0''$.

128. The axis of the lunar orbit is not at rest, but has a progressive motion, like that of the sun's orbit. This motion is $0^\circ\ 6'\ 41''$ in a day, or $40^\circ\ 41'\ 33''$ in a year; so that it makes an entire revolution, relatively to the fixed stars, in $3232^d.5807$, or in a little more than nine years.

The *Tropical* revolution of the perigee is shorter by $1^d.1056$.

The motions hitherto enumerated, are similar in the Moon and in the Sun. There are other inequalities peculiar to the moon.

129. The moon's longitude, calculated according to the laws of the elliptic motion, does not agree exactly with her true place, but requires to be corrected by an arch proportional

to the sine of double the angular distance of the moon from the sun, *minus* her mean anomaly.

a. If ☾ be the mean longitude of the moon, and ☉ that of the sun; the mean anomaly of the moon being x, this inequality is

$$(1^\circ\ 21'\ 5''.5) \times \sin(2(☾ - ☉) - x).$$

b. This inequality is called the *Evection*: it was discovered by PTOLEMY, and is, after the equation of the centre, the first of the lunar irregularities that was observed.

c. The argument of it, or $2(☾ - ☉) - x$, increases at the rate of 11° 18′ 59″, or 11°.3166 *per* day; so that its period is $\frac{360}{11.3166}$, or 31d.8119; that is, in the space of 31 days 19 hours 28 minutes nearly, the evection runs through all its changes, and is beginning to renew them in the same order.

d. At the new and full moon, or at the *Syzygies*, when ☾ — ☉ is either nothing or 180°, the argument is $-x$, which gives the evection negative, if x be less than 180°, and positive, if it be greater, contrary to what happens to the equation of the centre, which is therefore diminished in both cases. At the quadratures, the equation to the centre is increased by the evection. The evection appears, therefore, as an inequality in the equation of the centre, by which it increases when the moon is in the

the quadratures, and diminishes when the moon is in the syzygies.

130. Another irregularity in the moon's motion, which vanishes at the syzygies and the quadratures, and is greatest in the middle between them, or in the *octants*, is called the *Moon's Variation.* It depends on the difference of longitude of the sun and moon, or on their angular distance, and may be represented by the formula

$$(35' \, 42'') \sin 2 \, (☾ - ☉).$$

This inequality was discovered by TYCHO BRAHE'. Its period is $14^{d}.7653$, or half a lunar month.

131. The angular motion of the moon is also subject to a third irregularity, by which it is diminished when the sun approaches the perigee, and increased when he approaches the apogee. This is called the *Annual Equation:* it depends entirely on the time of the year, or mean anomaly of the sun, and is equal to

$$-(11' \, 11''.9) \sin \text{Mean Anom.} ☉.$$

TYCHO BRAHE' was acquainted with this inequality, and made an allowance for it in the equation of time. These are the only irregularities in the moon's motion, that were known previously to the theory of Gravitation. That theory not only explains

plains them, but has led us to distinguish many more, of which the accumulated effect was perceived, by the disagreement of the calculated with the observed places of the moon, though their laws were unknown. These will be considered under the head of Physical Astronomy.

Of the Moon's Revolution on her Axis.

132. The moon, as she revolves in her orbit about the earth, preserves always, at least nearly, the same face turned toward the earth, and therefore must revolve on her axis, in the same direction, and in the same time, that she revolves in her orbit.

a. The moon's surface, when viewed with a telescope, presents an object so strongly characterised, as to leave no doubt of its being always the same.

The accurate examination of the spots, which the tracing of the progress of the illumination has naturally induced, has discovered some apparent inequalities in the moon's revolution on her axis, to which the name of *Libration* has been given. They are optical appearances, and argue no real inequality in the moon's rotation.

133. If

133. If the angular velocity with which the moon revolves on her axis is uniform, that with which she revolves in her orbit, being, as we have seen, alternately accelerated and retarded, small segments on the east and west sides ought alternately to come in sight and to disappear. This is conformable to observation.

Thus there is produced in the orb of the moon, an apparent libration backward and forward, called her *Libration in Longitude.*

134. If the axis on which the moon turns, were perpendicular to the plane of her orbit, or if her equator coincided with that plane, we should perceive no other libration than that which has now been described. But in fact the spots on the north and south of the plane of the moon's orbit, alternately advance toward the middle of the disk, and recede from it by a very small quantity.

This is called the *Libration in Latitude*, and shews that the moon's axis it not exactly, though nearly, perpendicular to the plane of her orbit.

135. A third libration arises from this, that the point to which the moon presents always the same face, is the centre of the earth, round which

which the spectator describes a circle parallel to the equator, in the course of the diurnal revolution of the earth.

Hence, when the moon rises, a spectator sees some spots toward the upper limb of the moon, which he would not see at the centre. As the moon becomes more elevated, these points approach more to the upper edge, while others, on the inferior limb, come in sight. This order is reversed as the moon descends to the horizon. This is called the *Diurnal* or *Parallactic Libration.*

136. From an attentive observation of the lunar spots, it has been found, that the equator of the moon is inclined to the plane of the ecliptic, at an angle of 1° 30′; and that the line in which its plane cuts the plane of the ecliptic, is parallel to the line of the nodes, or to that in which the moon's orbit cuts the same plane.

Biot, tom. II. p. 411. § 72.

a. Suppose three planes to pass through the centre of the moon, one representing the equator of the moon, another the plane of her orbit, and let the third be parallel to the ecliptic. This last will lie between the two others, and will intersect them in the same line in which they intersect one another.

With

With the first, it will make an angle of 1° 30′; and with the second an angle of 5° 9′.

This curious fact was observed by CASSINI. LA GRANGE has explained it from the theory of gravity.

SECT. VIII.

ECLIPSES OF THE SUN AND MOON.

137. THE Sun and Moon are both subject to occasional obscurations, most frequently partial, but sometimes also total, which are called *Eclipses.*

138. The Eclipses of either of these luminaries happen only when the moon is near her nodes; that is, either in the plane of the ecliptic or near it. Those of the sun happen only at new moon, or when the moon is in conjunction with the sun: the eclipses of the moon, again, happen at the full moon, or when the moon is in opposition to the sun. The three bodies, then, the Sun, the Earth and Moon, are always nearly in the same straight line when an eclipse happens; and conversely, when these three

three bodies are in a straight line, or not distant from it by more than certain limits, an eclipse always takes place.

139. Hence it is evident, that an eclipse happens, in consequenee of one of the two opaque bodies, the Earth and the Moon, being so placed as to prevent the Sun's light from falling on the other.

The interposition of the moon between the sun and the earth, produces an Eclipse of the Sun; and the interposition of the earth between the moon and the sun, so that its shadow falls on the moon, or on any part of the moon, produces an Eclipse of the Moon. The whole of the phenomena of eclipses admit of explanation, on these principles.

140. As the return of eclipses must depend on the return of the line of the syzygies to the line of the nodes; and as the mean angular motion of these lines is known, the periods at which eclipses would return, were there no irregularity in the motions of the earth and of the moon, may be easily calculated.

a. The time of a lunation, or of one revolution of the line of the syzygies, is $29^{d}.530588$, as already stated; and a revolution of the line of the nodes, relatively to the sun, is $346^{d}.61963$. If, by the method of continued fractions, we seek for smaller numbers,

numbers, that may nearly express this ratio, we shall find 19 and 223; so that after 223 lunations, the node has nearly performed 19 revolutions. In 223 lunations, therefore, or 18 Julian years 10 days and 7 hours, the sun, the moon, and the node are nearly in the same position with respect to one another; and the series of eclipses returns nearly in the same order.

This period is thought to be the SAROS of the Chaldean astronomers; and their predictions of eclipses were probably founded on it. It is particularly mentioned by PLINY. A period of 521 Julian years, is considerably more exact. LA LANDE, *Astron.* § 1503.

b. The mean motions of the lines of the syzygies, and of the nodes, are also used in another way in the calculation of eclipses. Though the mean places of those lines are different from the true, and though it be on the latter that the phenomena of eclipses depend, it is useful to have the mean places, in order to know whether the circumstances are such, that an eclipse can possibly happen or not at a given new or full moon. The mean motions in astronomical tables, afford the means of performing this calculation, which is also much abridged by a table of what are called the *Epacts,* that is, of the ages of the moon, (reckoning from the last mean conjunction, and supposing her motion uniform), at the beginning of every year. LA LANDE, *Astron.* § 1732.; FERGUSSON, *Astron.*

Eclipses

Eclipses of the Moon.

141. The length of the earth's shadow varies, according to the distance of the sun and earth, between the limits of 212.896 and 220.238 semidiameters of the earth; its mean length being 216.531.

Half the angle of the cone formed by the shadow of the earth, = semid. ☉ — parallax ☉; and therefore if r be the radius of the earth, S being the apparent semidiameter, and p the horizontal parallax of the sun, the length of the shadow reckoned from

the earth's centre $= \dfrac{r}{\sin(S - p)}$.

142. Hence half the angle subtended at the earth's centre, by the section of the shadow at the distance of the moon, is $P - S + p$.

P is the horizontal parallax of the moon, S and p as above.

The apparent semidiameter of the shadow, may therefore vary from 37′ 42″ to 45′.52.

See Woodhouse, *Astron.* p. 340.

As

As the diameter of the shadow may be greater than three times the diameter of the moon, the moon may be totally eclipsed for as long a time as she takes to describe twice her own diameter, that is, for about two hours.

143. The phenomana of a lunar eclipse may be understood, by conceiving two circles of given magnitudes, the disk of the moon, and the section of the earth's shadow, as moving in the same plane with given velocities, and given directions; the moments when they touch, when they cease to touch, and when their centres are nearest to one another, determine the beginning, the greatest obscuration, and the end of the eclipse.

As these circles move in the same plane, and are in contact with one another, the place of the spectator makes no difference as to the preceding phenomena, which must happen at the same instant of absolute time to all observers, in whatever part of space they are situated. Hence the phenomena of a lunar eclipse are in themselves far simpler, and more easily calculated, than those of a solar eclipse, where the body that obscures the other is distant from it, so that their apparent places are much affected by the place and the motion of the observer.

144. Considering

144. Considering the small arches moved over by the moon, and the section of the shadow, during the time of an eclipse, as straight lines given in position, and described by two points, viz. the centre of the moon's disk, aud the centre of the section of the shadow, moving with given velocities, the determination of the phenomena of a lunar eclipse is reduced to the solution of a geometrical problem.

Let ST and MP, (fig. 14.), represent the portions of the ecliptic, and of the moon's orbit traversed, during the time of a lunar eclipse, the first by the centre of the section of the shadow, and the second by the centre of the moon, considered as straight lines. Let S be the centre of the earth's shadow, and M the centre of the moon at the instant of the *opposition*, and let S′ and M′ be any other cotemporary positions of these centres; SS′ and MM′ being taken in opposite directions. Draw S′N parallel to SM; and join M′N. Let t be the time in which SS′ and MM′ have been moved over, reckoned from the moment of the opposition, in hours and decimals of an hour; let the horary motion of the moon in longitude be m, so that $\mathrm{MO} = m\,t$; her horary motion in latitude λ, so that $\mathrm{M'O} = \lambda\,t$; and let the horary motion of the sun or of the shadow be n, then $\mathrm{SS'} = \mathrm{MN} = n\,t$.

Also

Also, tan. NM′O $= \frac{m+n}{\lambda}$, so that the angle NM′O

$=$ S′NM′ is given ; let this angle $= \beta$,

then NM′ $= t\sqrt{(n+m)^2+\lambda^2} = r\,t$,

r being $= \sqrt{(n+m)^2+\lambda^2}$. Hence, if SM $= a$,

$$S'M' = \sqrt{r^2t^2 - 2\,a\,r\,t \times \cos\beta + a^2};$$

and if the sum of the semidiameter of the moon and shadow be called s, we have the equation $r^2\,t^2 - 2\,a\,r\,t\cos\beta + a^2 = s^2$, for determining t.

Hence $t = \frac{a}{r} \pm \frac{\sqrt{s^2 - a^2\sin\beta^2}}{r}$, or

$$= \frac{a \pm \sqrt{s^2 - a^2\sin\beta^2}}{r}.$$

These two values of t give the time of the beginning and end of the eclipse.

The least distance of the centres, is found by substituting $\frac{a}{r}$ for t in the formula

$$S'M' = \sqrt{r^2\,t^2 - 2\,a\,r\,t \times \cos\beta + a^2},$$

S′M′

S'M' is then $= \sqrt{a^2 - 2\,a^2 \cos\beta + a^2} =$

$a\sqrt{1 - 2\cos\beta + 1} = 2\,a \sin\frac{1}{2}\beta = 2\,a$ multiplied into the sine of half the apparent inclination of the moon's orbit.

When $s^2 = a^2 \sin\beta^2$, or $s = a \sin\beta$, there can be no more than a contact of the moon and shade, which will happen at the instant expressed by by $\frac{a}{r}$.

When $s = a \sin\beta$, $\sin\beta = \frac{s}{a}$, and if $\sin\beta > \frac{s}{a}$, there can be no eclipse, as $\sin\beta = \frac{\lambda}{r}$; therefore, unless $\frac{\lambda}{r}$ be less than $\frac{s}{a}$, there can be no eclipse. Now,

$r = m + n + \frac{\lambda^2}{2\,(m+n)}$ nearly; so that

$$\frac{\lambda}{m + n + \frac{\lambda^2}{2\,(m+n)}}$$

must be less than $\frac{s}{a}$, otherwise there can be no eclipse of the moon.

It will serve as a limit, to consider whether $\frac{\lambda}{m+n}$ is

less

less than $\frac{s}{a}$; or whether the horizontal motion of the moon in latitude, have to the sum of the horary motions of the moon and sun in longitude, a less ratio than the sum of the semidiameters of the moon and shadow has to the latitude of the moon at the time of the conjunction.

If $a = 0$, $t = \frac{a \pm s}{r}$.

When the moon merely touches the shadow, that is, when the nearest approach of the moon to the shadow is just equal to the sum of the semidiameters of the moon and of the section of the shadow, we have what is called an *Appulse*.

To derive as much advantage from the knowledge of the ecliptic limits as possible, it is necessary to observe, that when the mean opposition is 12° 36′ distant from the node, there can be no eclipse; and that when it is less than 9° distant from it, there must be an eclipse. Between these limits 12° 36′ and 9°, the matter is uncertain, and must be decided by the calculation of the true place of the moon, &c. These are the limits given by De Lambre; Woodhouse's *Astron.* p. 342. The limits are differently assigned by other writers, Vince, *Astron.* vol. i. § 545.

145. The moon seldom disappears entirely in lunar eclipses; even the spots may be distinguished

guished through the shade; the nearer the moon happens to be to the earth, or the farther she is from the *apex* of the shadow, the darkness is the greater.

The light that, by the refraction of the earth's atmosphere, is made to enter within the limits of the conical shadow, is no doubt the cause of this phenomenon.

In some instances the moon has disappeared entirely, as in that mentioned by KEPLER in June 1620. HEVELIUS has taken notice of another, where the moon could not be seen even with a telescope, though the night was remarkably clear.

146. As an eclipse of the moon happens at the same instant of absolute time to all observers, it is one of the phenomena, from the observation of which the longitudes of places may be most directly inferred, § 67.

On account of the ill-defined boundary of the shadow, this method of ascertaining the longitude does not admit of great precision. It is difficult to determine the beginning or end to less than a minute of time.

The arrival of the boundary of the shadow, at the different spots, can be more accurately ascertained than the beginning or end; and, therefore, as many observations of that kind should be made as possible.

When several such observations, made under two different meridians, are compared, the mean may furnish a tolerably exact determination of the difference of longitude.

The comparison of the beginning or end, with calculations previously made, may also serve for finding the longitude, and may be useful for that purpose at sea.

Solar Eclipses.

147. The length of the moon's shadow is less than that of the earth, in the same ratio that the diameter of the moon is less than the diameter of the earth; that is, in the ratio of 1 to 3.562.

a. Hence, when the Earth is in the aphelion, the length of the Moon's shadow is 59.730; and if the moon is in the perigee, its distance from the earth is only 55.902; so that the shadow may reach the earth, and a total eclipse may take place. But if the moon were in her apogee, when her distance is 63.862, the shadow could not reach the earth, and the eclipse could not any where be total.

b. When the earth is in the perihelion, the length of the moon's shadow is 57.76; and if at the same time the moon be in the perigee, or indeed nearer than her mean distance, a total eclipse may happen.

c. The moon's mean motion about the centre of the earth is 33′ in an hour; and the shadow of the moon, therefore, traverses the surface of the earth when it falls on the surface perpendicularly, with a velocity of about 380 miles in a minute. When the shadow falls obliquely, its velocity appears greater in the inverse ratio of the sine of the obliquity.

Relatively to a point on the earth's surface, the shadow may go much faster than this, as its motion may be in an opposite direction to the diurnal rotation.

The duration of a total eclipse, in any given place, cannot exceed 7′ 58″. La Lande, § 1777.

An Annular Eclipse, or one where the sun's disk appears like a ring all round the moon, may last 12m. 24s.

To have a partial eclipse of the sun, it is not requisite that the shadow should reach the earth; it is sufficient that the distance of the centres of the sun and moon be less than the sum of their apparent semidiameters.

148. When

148. When an opaque body is opposed to one which is luminous, there is a certain space behind the former body, from which the latter is only partially visible. This space is called the *Penumbra* of the opaque body.

a. If the bodies are spherical, the penumbra is a cone, having for its angle the line joining the centres of the two bodies; and for its vertex, the point in that line where tangents to the opposite sides of the two bodies intersect one another.

b. Half the angle of the conical penumbra of the moon, is equal to the apparent semidiameter of the sun, *plus* the angle which the moon's semidiameter subtends at the sun.

149. To conceive the phenomena of a solar eclipse in general, we may consider the section of the moon's penumbra as advancing on the earth from the west, and as being viewed by an observer in the moon, in the same manner that an eclipse of the moon is viewed by an observer on the earth.

The observer might be placed any where, providing he saw the whole of that side of the earth which is turned to the sun; but it is most convenient to suppose him in the moon, that the arches, compa-

red

red with one another, may belong to circles of the same radius.

150. The penumbra in a solar eclipse, if viewed from the moon, would subtend an angle equal to the sum of the diameters of the sun and moon, as seen from the earth.

When the penumbra just touches the disk of the earth, the distance of their centres is equal to half the angle subtended by the section of the penumbra, *plus* the moon's horizontal parallax; that is, = semidiam. ☽ + semid. ☉ + Hor. Par. ☽.

If the least distance of the centre of the sun and moon, (the same with the least distance of the centres of the earth and of the penumbra), is less than this, there can be no eclipse.

The greatest value of the above angle is about 1° 34′ 27″. Supposing this to be the distance at the time of the ecliptic conjunction, we get the distance from the node 17° 21′ 27″. If the conjunction happens nearer to the node than this, there may be an eclipse. If it be more distant there can be none.

The limits of solar eclipses being greater than of lunar, there are more eclipses of the sun than of the moon,

moon, and that nearly in the ratio of 3 to 2; but fewer eclipses of the sun are observed in any given place than of the moon, as a lunar eclipse is visible to a whole hemisphere; but a solar only to a part.

151. The general phenomena of the eclipse being calculuted to the time of a given meridian, the phenomena, as they will be observed at any particular place, may also be determined, by calculating the altitudes of the sun and moon, and the effects of their parallaxes for different instants of time, and then employing the method of interpolation, to determine the time of the beginning and end, and the quantity of the greatest obscuration.

a. Let the places of the sun and moon be found for an instant, far from the beginning of the eclipse, and from thence let their altitudes for the given place be computed, as also the effects of parallax in longitude and latitude.

Let the difference of the apparent longitude of the two bodies thus found be called , and the latitude of the moon λ. If the sun's parallax is included, let his parallax in latitude be applied to the moon, taking notice, whether it increases or diminishes the difference of latitude. Then ACB, fig. 15. being an arch of the ecliptic, A the place of the sun, D of the moon, as just computed, P the pole of the ecliptic, and PDC a circle of longitude, AC $= \delta$,

$= \delta$, and CD $= \lambda$, and AD the distance of the centres, which we may call $y = \sqrt{\delta^2 + \lambda^2}$ because the triangle ACD may be regarded as rectilineal. In practice, y may be found by a construction; or, if great accuracy is required, we may compute y from the trigonometrical formula, $\cos y = \cos \delta \times \cos \lambda$.

b. Let similar calculations be made for other two instants, separated by equal intervals of time m; so that one may be near the middle, and another near the end of the eclipse. Let the distances of the centres found for these times be A, A′, A″; let the differences of these distances be D and D′; and let the second difference, or $D - D' = \Delta$.

Then if y be the distance of the centres for any time t, reckoned from the instant for which the first computation is made, $y =$

$$A + \frac{D - \frac{1}{2}\Delta}{m} \cdot t + \frac{\Delta}{2\,m^2} \cdot t^2.$$

The distance of the centres is thus expressed in terms of the time, and from this equation the time of the beginning and end of the eclipse, and the quantity of greatest obscuration, may be determined.

c. The time of the greatest obscuration is $=$

$$\frac{m\,(D - \frac{1}{2}\Delta)}{D};$$

and this being substituted for t, the value of y will give the nearest approach of the centres.

d. This

d. This is only one of many methods that have been contrived for the calculation of eclipses. La Lande, *Astron.* tom. II. § 1860, &c.

e. The geometrical construction, by means of a projection, is sufficiently accurate for the prediction of eclipses. The most simple is that which supposes the observer to be placed in the sun, and to see the path which any place on the earth's surface describes in consequence of the diurnal motion, projected into an ellipsis on the plane of the earth's disk, while the path of the moon's shadow is projected into a straight line on the same disk. La Caille, *Astron.* Vince, § 575.

f. A very great addition to these methods of calculation was made by La Caille, viz., that by which a geographical representation is given of the path of the shadow, the quantities and the times of obscuration, &c. for all places of the earth. La Caille, *Astron.* 1166. La Lande, *Astron.* § 1911.

The construction of such a map may be made with sufficient accuracy by means of a celestial globe.

Number

Number of Eclipses.

a. In the space of eighteen years, there are usually about 70 eclipses, 29 of the moon, and 41 of the sun. These numbers are nearly in the proportion of 2 to 3.

b. Seven is the greatest number of eclipses that can happen in a year, and two the least.

c. If there are seven, five must be of the sun, and two of the moon. If there are only two, they must be both of the sun; for in every year there are at least two eclipses of the sun.

d. There can never be more than three eclipses of the moon in a year; and in some years there are none at all.

e. Though the number of solar eclipses is greater than of lunar in the ratio of 3 to 2, yet more lunar than solar eclipses are visible in any particular place, because a lunar eclipse is visible to an entire hemisphere, and a solar is only visible to a part. VINCE's *Astron.* § 588.

f. Central Eclipses are comparatively rare phenomena; for though there are about 28 such eclipses in every *cycle* of eighteen years, yet the space over

which

which every one of them appears to be central, is a narrow belt, perhaps a mere mathematical line, traced across the enlightened hemisphere of the earth.

g. A central eclipse is *annular*, when the angle subtended by the sun's diameter is greater than that subtended by the moon's; it is total for an instant, or *sine morâ*, when these angles are equal; and it is total for a portion of time that can never exceed eight minutes, when the angle subtended by the moon's diameter is greater than that subtended by the sun's.

h. A central eclipse observed at London in April 1715, is described by Dr HALLEY. The darkness for a few minutes was so entire, that the stars became visible. Though the disk of the sun was wholly covered by the moon, a luminous ring of a faint pearly light surrounded the body of the moon the whole time. Its breadth was about a tenth of the moon's diameter. The longest time that the obscuration lasted any where in Britain, was about 3^{m} 57^{s}. *Phil. Trans.* vol. 29. p. 295., &c. VINCE, *Astron.* vol. 1. § 585.

Occultation

Occultation of Stars.

152. The same method used in calculating eclipses, may be applied to compute the occultation of a fixed star by the moon; only, when the moon is distant from the ecliptic, the base of the right-angled triangle in the former construction, must not be supposed equal to the difference of longitude, but to that difference multiplied into the sine of the star's distance, from the pole of the ecliptic, or into the cosine of its latitude.

If S (fig. 16.) be the star, D the moon, DE the difference of latitude, SE is not to be taken as equal to AC, the difference of longitude, but as equal to $AC \times \sin SP = \text{dif. lon.} \times \cos \lambda$.

The distance $SD = \sqrt{SE^2 + DE^2}$, being thus found, the rest of the computation is as before.

If at the time of the mean conjunction of the moon and a star, that is, when the moon's mean longitude is the same with the longitude of the star, their difference of latitude exceed 1° 37′, there can be no occultation, but if the difference be less than 51′, there must be an occultation somewhere on the face of the earth. Between these limits there is a doubt, which can only be removed by the calculation of the moon's true place. VINCE, vol. I. § 591:

All

All the stars of which the latitude is less than 4° 32′ may suffer occultations by the moon, in any part of the earth. Dr BREWSTER has given a catalogue of the most remarkable stars subject to occultations of the moon. *Edinburgh Encyclopædia*, art. *Astronomy*, vol. II. p. 668.

SECT. IX.

OF THE PLANETS.

153. IT was said, (§ 3. *a*.) that beside the sun and moon, ten of the stars have motions *eastward*, peculiar to themselves. They are called Planets, and are distinguished by particular names, which, taken in the order of the celerity of their motions, are *Mercury, Venus, Mars, Vesta, Juno, Ceres, Pallas, Jupiter, Saturn, Uranus*.

a. The first two perform their revolutions in the heavens in less than a year, and are called *inferior* Planets; the rest have their period greater than a year, and are called *superior*.

b. Five of the Planets, Mercury, Venus, Mars, Jupiter, and Saturn, are very conspicuous, and have been known from immemorial time. PLINY says of them: "Suus quidem cuique color est; Saturno candidus, Jovi clarus, Marti igneus, Lucifero candens, Vesperi refulgens, Mercurio radians; Soli cùm

cùm oritur ardens, posteà radians." *Hist. Nat.* lib. II. cap. 16.

By Lucifer is understood Venus, when seen in the morning before sun-rise. By Vesper, the same planet seen in the evening after sun-set.

c. The other five planets are visible only through the telescope, and have been lately discovered; Uranus by HERSCHEL, in 1781; Ceres by PIAZZI, in 1801; Pallas by OLBERS, in 1802; Juno by HARDING, in 1803; Vesta by OLBERS, in 1807.

d. The planets have also particular characters, by which they are distinguisded; these, in the order in which they have been enumerated, are,

☿ ♀ ♂ ⚶ ⚵ ⚳ ⚴ ♃ ♄ ♅.

It is best to begin with the inferior planets, and with Venus, as that of which the phenomena are most easily observed.

154. *Venus*, the most brilliant of the planets, always accompanies the sun, never receding from him more than 45°, and becoming, as she is on the east or west side, alternately the Evening or the Morning Star.

a. Venus is the only planet mentioned in the Sacred Writings, and in the most ancient poets, such as HESIOD and HOMER.

b. The

b. The Evening and Morning Star, or the Hesperus and Phosphorus of the Greeks, were at first supposed to be different. The discovery that they are the same is ascribed to PYTHAGORAS.

155. This planet, when an evening star, and at her greatest distance from the sun, or at what is called her Greatest Elongation, appears, through the telescope, to have a semicircular disk, like the moon in the last quarter, with its convexity turned to the west. From that time, during her approach to the sun, her splendour increases for a while, though the quantity of the illuminated disk diminishes, like the moon in the wane; and at the same time, her diameter, measured by the distance of the horns, increases.

a. At the time of her greatest elongation, Venus is stationary with respect to the sun, or has the same motion in longitude. After that, her motion in longitude becomes slower than the sun's, and she comes nearer to the sun, as just remarked. At a certain point she becomes stationary with respect to the fixed stars, having no motion in longitude. After that, her motion becomes retrogade in respect of the fixed stars, and is directed westward.

b. Venus at last approaches the sun, so as to be lost in his light; and after some time, appears on the

west

west side, and is seen in the morning, before the sun rises.

c. Though Venus in general is not visible at the time of her conjunction with the sun, she has sometimes been seen as a dark spot passing over the body of the sun. This is the phenomenon called the *Transit of Venus*. Her diameter is then greatest, and measures nearly one minute.

156. As Venus proceeds to the westward, her disk is seen as a crescent continually increasing, at the same time that the diameter is diminishing. At the elongation of 45°, the disk is again a semicircle; and from thence it increases, while the distance from the sun diminishes, till the planet is lost in the sun's rays; her orb being almost a circle, but its diameter not more than one-sixth of what it was at the former conjunction.

a. The conjunction, which is preceded by the approach to a full orb, or that which follows the western elongation of Venus, is called the *superior conjunction*, as she is then farthest from the earth. The other the *inferior*.

b. The time of the greatest elongation of Venus is about sixty-nine days before or after the inferior conjunction, when she is between 39° and 40° distant from the sun, and comes to the meridian 1^h 38^m either before or after noon; her disk is then like

like that of the moon five days before or after the conjunction. La Lande, § 1147.

157. After the superior conjunction, the orb of Venus increases in magnitude; but the enlightened part diminishes, just reversing the former order, till she arrive at her greatest eastern elongation; after which the phenomena are repeated, as already mentioned; and the period which circumscribes all those changes, or the time from one conjunction to the next conjunction of the same sort, is, in its mean quantity, 584 days.

a. This is called the *Synodical Revolution of Venus.* She is retrograde with respect to the fixed stars during forty-two days of that period.

158. Hence it is evident, that the orbit of Venus surrounds the Sun, but excludes the Earth; and that her motion, with respect to the sun, may be equable, notwithstanding the variety of appearances it puts on to a spectator in the earth.

a. If a be the distance of the earth from the sun, and x that of Venus from the sun, $a + x$ will be its distance from the earth, at the superior conjunction, and $a - x$ at the inferior; and by § 156. $a + x : a - x :: 6 : 1$; so that $2a : 2x :: 7 : 5$, or as 10 to 7. If the earth's distance from the sun be called 10, that of Venus is 7; and this serves as

a

a first approximation to the distance of Venus, supposing her orbit a circle, with its centre in the sun.

b. From the synodical revolution of 584 days, the periodic time of Venus is found nearly equal to 224 days. For if v be the angular velocity of the planet in her orbit, or her diurnal motion, that of the earth being 59′ 8″.3; then the arch described by the sun, during a synodical revolution of Venus is (59′ 8″.3) 584; and that described by Venus in the same time is 360° + (59′ 8″.3) 584; therefore the diurnal motion of Venus in her orbit

$$= \frac{360° + (59' \; 8''.3) \; 584}{584} = 59' \; 8''.3 + \frac{360°}{584}$$

= 1° 35′ 56″; and the periodic time =

$$\frac{360°}{1° \; 35'.56} = 224 \text{ days, nearly.}$$

159. *Mercury*, like Venus, accompanies the sun, and never recedes from him farther than 28°. The greatest apparent diameter is 11″.275, the least 5″.025.

a. Hence Mercury describes an orbit round the sun, at the mean distance, (computed as in the case of Venus), of 3.7 of the parts whereof the Earth's distance from the sun is 10.

b. The synodic revolution of Mercury is 115d.877 nearly.

1

The

The diurnal motion in his orbit 4° 5′ 32″.

The periodic time 87^{d}.97 nearly.

160. The points in which a planet has no latitude, are called, as in the case of the Moon, the *Nodes of the Planet.* The Planet is then in the ecliptic, and one-half of its orbit lies on the north, the other on the south side of that plane.

a. The line of the nodes of every planet, or the common section of the plane of its orbit, with the plane of the ecliptic, passes through the Sun. This was discovered by KEPLER. See Dr SMALL's *Account of* KEPLER's *Discoveries*, p. 154.

b. The node through which the planet passes into the northern signs, is called its *Ascending Node;* that through which it passes into the southern, is called the *Descending Node.*

c. The *Heliocentric* place of a planet, is its place as it *would* be seen from the Sun: the *Geocentric*, as it *is* seen from the Earth.

161. When the Earth is in the line of a planet's nodes, or, which is the same, when the Sun is seen from the Earth in that line, if the planet's elongation from the Sun, and its geocentric latitude be observed, the inclination of the orbit may be found; for, the sine of the elon-

gation is to the radius, as the tangent of the geocentric latitude to the tangent of the inclination.

La Lande, § 1358. Gregory's *Astron.* Book III. prop. 20.

If the planet be 90° distant from the Sun, the latitude observed is just equal to the inclination. Kepler made use of this last method for determining the inclination of the orbit of Mars.

162. If an inferior planet, at the inferior and superior conjunctions, or a superior planet at the opposition and conjunction, be also 90° distant from the node; from the observation of its geocentric place, the inclination of the orbit being known, the ratio of the planet's distance from the Sun, to the Earth's distance from the Sun, may be found.

In the annexed figure, (fig. 17.), let S be the Sun, VV′ the orbit of Venus, E the Earth; then joining EV, EV′, and drawing the perpendiculars VD, V′D′, if $ES=a$, $SV=b$, and the angle $ESV=I$, $SD=b\cos I=SD'$; so that $ED=a-b\cos I$, and $ED'=a+b\cos I$. If the angle $VES=\lambda$, and $V'ES=\lambda'$, then because $VD=V'D'$,

tan

$\tan \lambda : \tan \lambda' :: a + b \cos I : a - b \cos I$. And hence $\frac{b}{a} = \frac{\tan \lambda - \tan \lambda'}{\cos I (\tan \lambda + \tan \lambda')}$.

This affords a first approximation to the distance of the planet, supposing its orbit to be circular.

163. The orbits of the superior planets, include both the Sun and the Earth within them; and the Sun is nearly in the centre of each.

a. The superior planets are seen in opposition to the Sun as well as in conjunction; and in the latter situation, the angle subtended by the planet is much less than in the former. The angle, for example, under which the diameter of Mars is seen in opposition to the Sun, is 18″.972; and in conjunction 3″.96, only a fifth part of the former. Hence, reasoning as before, a being the distance of the Earth from the Sun, and b the distance of Mars from the same, $b + a : b - a :: 5 : 1$, and $2b : 2a :: 6 : 4$, or $b : a :: 3 : 2$.

b. When the oppositions of Mars are observed in different places of the heavens, and when, of consequence, Mars is in different points of his orbit, the ratio of b to a is still found nearly the same; hence, as we know that a does not vary much, the variation of b must be inconsiderable, and the Sun must therefore

therefore be nearly in the centre of the orbit of Mars. The same is true of Jupiter, Saturn, &c.

c. Mars appears with his disk perfectly round, both at the opposition and the conjunction. In the intermediate positions, he is found to want something of perfect rotundity on the side turned from the Sun.

Observations made at the opposition of a planet, and, when possible, near the conjunction, are valuable, because the planet is then seen from the Earth in the same place it would be seen in from the Sun; that is, the geocentric and heliocentric places, either coincide, or differ exactly by 180 degrees.

164. The superior planets are not always progressive, or do not always move from west to east; but, like the inferior, they become stationary, and also retrograde. They are progressive at the conjunction, and for a considerable distance on each side of it; they are retrograde at the opposition, and at certain points between, are stationary, passing gradually from one of these states to another.

a. Thus Mars, when he emerges from the Sun's rays, a few days after the conjunction, and is seen to

rise

rise some minutes before the Sun, is found to be progressive, that is, his right ascension, if observed from day to day, continually increases. The right ascension of the Sun, however, increases faster, so that Mars recedes from the Sun toward the west, though his real motion, with respect to the fixed stars, is toward the east. The motion, however, is continually growing slower, and at the end of a year nearly, the angular distance from the Sun being then between 136° and 137°, the planet becomes stationary, and is without any sensible motion for a few days.

The motion then becomes retrograde, or toward the west, and continues so till Mars be 180° distant from the Sun, or in opposition, so as to be on the meridian at midnight. The retrograde motion is then swiftest; it afterwards becomes gradually slower, and ceases altogether when the planet has again come to be between 136° and 137° distant from the sun on the other side. The motion becomes then progressive, and continues so till the conjunction, and beyond it, in the manner just described.

b. The same is true, with certain variations, of all the other planets. The longer the period, and the more distant the planet from the Sun, the longer the time, and the smaller the arc of retrogradation.

For

For the six principal planets, these arcs are exhibited in the annexed Table, according to their mean quantities. See LA LANDE, § 1192. Also BIOT, § 97. vol. III.

TABLE.

	Elongation, when stationary.	Arch of Retrogradation.	Time of Retrogradation.	Synodic Revolution.
Mercury,	18° 00′	13° 30′	23 days.	118 days
Venus,	28 48	16 12	42	584
Mars,	136 48	16 12	73	780
Jupiter,	115 12	9 54	121	399
Saturn,	108 54	6 18	139	378
Uranus,	103 30	3 36	151	370

165. The apparent motion of an object is affected by the motion of the spectator; and if there is not a certainty that he is at rest, when the motion of the former appears extremely irregular, it is natural to inquire, whether any motion

tion that can reasonably be ascribed to the latter will explain the irregularities observed.

In the present case, there is not a certainty that the spectator, (or the Earth), is at rest: the want of any information of his motion from terrestrial objects affords none, as, according to a principle already explained, the motions of bodies among one another, are nowise affected by any motion which is common to them all.

166. When a spectator moves, without being sensible of it, he necessarily transfers his own motion to the objects around him, estimating it in a direction opposite to that in which he has actually moved.

Suppose an object to move from A to B, (fig. 18.), while an observer, unconscious of his own motion, is carried from C to D. From B draw BF equal and parallel to CD, but extending from B the opposite way that CD does from C, and join AF; AF will be the apparent path of the body, and F its apparent place, at the time when the spectator is really in D, and the body in B.

If the velocity of the observer had been the same, but in an opposite direction, CD′, BF′ being drawn = CD′, but in the opposite direction, the apparent motion of A would have been in the line AF′

This

This is evident, because CF′ is equal to D′B, and makes the same angle with CD.

Hence it is evident, that an object, without being really at rest, may be apparently so, if the observer is in motion, and may even acquire an apparent motion, in a direction contrary to its real. Thus also the observer and the object both moving with perfect regularity, both describing concentric circles, for example, with uniform velocities, and directed the same way, the one may become stationary in respect of the other, and even acquire a motion in an opposite direction.

167. Suppose d to be the distance of a planet (imagining it to describe a circle round the sun) from the sun, or the radius of its orbit, and e the elongation from the sun, at which it appears stationary, the radius of the circle in which the observer must move, in order to see the planet stationary at that elongation, being called x, is found from the equation, $x^2 + d\,x = d^2 \cot^2 e$;

$$\text{or } x = -\frac{1}{2} d + \frac{d}{2}\sqrt{1 + 4 \cot^2 e}.$$

a. This follows from what KEIL has demonstrated in his Astronomy, Sect. 27.; LEMONIER, *Institutions Astronomiques*,

Astronomiques, p. 585. La Lande has simplified the demonstration, § 1188. See also Biot, vol. iii. p. 181., Note.

b. If, by help of the above formula, we inquire, supposing the Earth to revolve in a year round the Sun, and Mars in 686.98 days, what must be the ratio of the distance of the Earth to the distance of Mars from the Sun, in order that the latter may be stationary at the elongation of 136°? we shall find it to be that of 1 to 1.52, which is the ratio deduced from other phenomena.

c. Not only are the stations thus explained, but the extent of the arches of progression and retrogradation also. This coincidence affords a strong presumption in favour of the system of the Earth's motion, or that which, from the name of its discoverer, is called the *Copernican System*.

d. The same holds of Jupiter, Saturn, Uranus. The same motion of the Earth, and the same distance from the Sun, will account for the phenomena in all the cases; so that whatever probability there is, from the phenomena of one planet, in favour of the Earth's motion, the same is increased in a quadruplicate ratio, from considering the phenomena of all these four superior planets.

e. The two inferior planets, give a similar increase of evidence. The four new planets are not taken into

into account, as the times of their stations, &c. may not yet have been sufficiently determined by observation.

f. On the strength of this evidence, we shall assume the motion of the Earth as a fact, and try whether it is consistent with the other phenomena of the planetary motions.

Orbits of the Planets.

168. If a planet be observed twice in the same node, the node in the interval being supposed to remain fixed, the position of the line of the nodes may be determined, and also the distance of the planet from the sun at the times of observation.

a. Let a superior planet be observed in its node N, (fig. 19.), from the Earth at E, and after the planet has made an entire revolution, and returned to the point N, let the Earth be at E′. Then, from the time, and the construction of the Earth's orbit, EE′ is given, and the angles SEE′, SE′E. But the angles SEN, SE′N, are known by observation; therefore the angles EE′N, E′EN, as also the

the base EE′ are given, and therefore the sides EN, E′N, are also given. Hence, from either of the triangles ESN, E′SN, SN is found, and also the angle ESN, or the heliocentric place of the node. GREGORY's *Astronomy*, Book III. prop. 19. WOLFII, *Elem. Astron.* § 777. Thus the periodic time of the planet is determined, and also its mean motion.

b. When observations of this kind are made at a considerable distance of time from one another, it is found that the nodes of every planet have a slow motion retrograde, or in a direction contrary to the order of the signs.

169. The distance of a planet from the Sun, and its heliocentric place, or its longitude as seen from the Sun, may be determined by observations made at the time of its opposition to the Sun.

If E (fig. 20.) be the Earth, S the Sun, P a planet, O its place reduced to the ecliptic, SN the line of the nodes; the points S, E, and O are in the same straight line, because of the planet's opposition to the Sun, and the angle ESN, that is, OSN, is known from the last problem; therefore the angle PSO or PSE, the heliocentric latitude, is also given. But the angle PEO, the geocentric latitude, is given by observation, and the base ES, from the theory of the Earth's motion; therefore SP,

or

or the planet's distance from the Sun, is found. The position of PS, relatively to NS, is also thus determined; for, in the right-angled spherical triangle, of which the base is the arch that measures the angle OSN, and the perpendicular the arch which measures the angle PSO, the hypothenuse is the measure of the angle PSN, which the radius vector makes with the given line SN.

b. Thus also, EP, the planet's distance from the Earth, is found. If, then, by observations made a little before and after the opposition, the diurnal motion of the planet in its own orbit, relatively to E, be determined, the same may be found relatively to S, being to the other in the inverse ratio of SP to EP.

When many oppositions of a planet are thus observed, many different radii of the planetary orbits are determined, as well as the angular motions corresponding to them.

170. It appears, on laying down the radii determined as above, that the orbits of the planets are ellipses, having the Sun in their common focus; and that the angular motions of a planet round the Sun, are inversely as the squares of its distances from the Sun; so that the sectors described by the radius vector, are proportional to the time.

These

These two propositions, which have already been shewn to hold of the Earth's motion, are therefore common to the motions of all the planets. They were discovered by KEPLER, and were first found out by him, with infinite ingenuity and labour, when he was endeavouring to determine the orbit of *Mars*.

171. When the focus of an ellipse, and three points in its circumference, are given, the ellipse may be described; and hence the planetary orbits may be determined, that is, the axis, the eccentricities, and thence the equations to the centres, &c.

The application of this to find the three *Elements* of an orbit, the Eccentricity, the place of the Aphelion, and the Epoch, or radical mean place, for a given time, is in LA LANDE, § 1288, &c. Also VINCE's *Astronomy*, vol. I. § 257. See also NEUTONII *Prin. Math.* lib. 1mus, prop. 21. Schol.

172. When the mean distances of the planets are compared, and also their periodical times, it is found that the squares of the periodical times are as the cubes of the distances.

This great general fact was also discovered by KEPLER, and is the third of the laws that bear the name of that astronomer.

173. When

173. When the elements of the orbit are found from observation, at periods considerably distant from one another, the line of the apsides of each of the planets is discovered to have a slow motion forward.

LA LANDE, § 1309, &c.

The Elements of the Orbits of the different Planets are given in the annexed tables.

I.

	Inclination of the Orbit to the Ecliptic for 1801.	Long. ascending Node for 1801.	Secular motion of the Node.
Mercury,	7° 0′ 1″	46° 57′ 31″	—13′ 2″
Venus,	3 23 32	74 52 52	—31 10
Mars,	1 11 6	48 1 30	—38 48
Vesta,	13 3 29	171 6 38	
Juno,	7 8 46	103 1 0	
Ceres,	10 37 34	80 1 3	
Pallas,	34 37 8	172 32 36	
Jupiter,	1 18 43	98 25 34	—26 17
Saturn,	2 29 38	111 55 27	—37 54
Uranus,	0 46 26	72 51 14	—59 57

II.

II.

	Siderial Revolutions.	Mean Distances.	Eccentricity, the mean Distance being 1.
Mercury,	87d.9692	0.38709	0.20551
Venus,	224 .7008	0.72333	0.00685
The Earth,	365 .2564	1.	0.01685
Mars,	686 .9796	1.52369	0.09313
Vesta,	1590 .998	2.66716	0.25494
Juno,	1335 .205	2.373	0.09322
Ceres,	1681 .539	2.76740	0.07835
Pallas,	1681 .709	2.76759	0.24538
Jupiter,	4332 .5963	5.20279	0.04817
Saturn,	10758 .9698	9.53877	0.05617
Uranus,	30688 .7127	19.18330	0.04667

III.

III.

	Mean Longitude of the Planets, 1st Jan. 1801.	Mean Longitude of the Perihelion.	Secular Mot. of the Perihelion.
Mercury,	166° 0′ 48″	74° 21′ 46″	9′ 43″
Venus,	11 33 16	128 37 1	4 28
Earth,	100 39 10	99 30 5	19 39
Mars,	64 22 57	332 24 24	26 22
Vesta,	267 31 49	249 43 0	
Juno,	290 37 16	53 18 41	
Ceres,	264 51 34	146 39 39	
Pallas,	252 43 32	121 14 1	
Jupiter,	112 15 7	11 8 35	11 4
Saturn,	135 21 32	89 8 58	32 17
Uranus,	177 47 38	167 21 42	4 0

The first of the above Tables gives the inclination of the orbits, the position of the line of the nodes, and the secular motion of the nodes for all the planets. The sign *minus*, prefixed to the motions of the nodes, signifies that they are retrograde. The inclinations of the orbits of Vesta, Ceres and Pallas, are greater than those of the other planets; and the orbit of the last goes far beyond the *zodiac*.

1 The

The secular variations of the nodes for these three planets, and for Juno, have not yet been determined.

The Second Table gives the times of the siderial revolutions in days; the mean distances from the Sun, or the semi-transverse axes of the orbits, on the supposition that the mean distance of the Earth is 1; also the eccentricity of each orbit, supposing the semi-transverse of that orbit to be 1.

The Third Table gives the mean longitude of each planet for the 1st of January 1801, to the meridian of Greenwich; also the mean longitude of the perihelion, or lower apsis of the orbit, with the secular variation. See *Exposition du Systeme du Monde*, 3me ed. BIOT, tom. III. § 29. WOODHOUSE, *Ast.* p. 286.

Rotation of the Planets.

176. Four of the planets, Venus, Mars, Jupiter and Saturn, when examined with the telescope, appear to revolve on axes, in the same direction in which they revolve in their orbits; the axis of each remaining always parallel, or nearly parallel to itself.

a. This conclusion is derived from the motion of certain spots, which are distinguished, by the colour or intensity of their light, from the other parts of the planetary disk. In this way the time of rotation is also determined, as in the case of the Sun, (§ 117).

b. It is thus found, that Venus revolves in $23^h\,21^m\,9^s$, on an axis which makes a very small angle with the plane of the ecliptic. This was first observed by the elder CASSINI. LA LANDE, § 3341.

c. Mars revolves in 1 day 39 minutes, on an axis inclined at an angle of 59° 42′ to the ecliptic.

d. Jupiter revolves in $9^h\,56^m$, on an axis nearly perpendicular to the ecliptic.

e. Saturn revolves on his axis in $10^h\,16^m$.

f. Though

f. Though Mercury is too much immersed in the Sun's rays to allow points of unequal splendour to be observed on his surface, yet certain periodical inequalities, observed in the *Horns* of the disk, seem to indicate a revolution on an axis.

Of the other planets, Uranus is too distant, and Juno, Vesta, Ceres and Pallas, too small, to admit of any observation from which either a revolution on an axis, or the contrary, can be inferred.

177. Jupiter, Saturn and Mars, are compressed at the poles, or the extremities of the axes of revolution, in the same manner with the Earth.

a. The compression of Jupiter amounts to a fourteenth part of his longer diameter; that of Saturn to an eleventh.

b. Mars, according to the observations of Dr HERSCHEL, is compressed at the poles, so that his equatorial diameter is to his polar axis as 16 to 15 nearly. *Phil. Trans.* 1781, p. 134. and 1784, p. 259. LA LANDE, § 3343.

178. Some appearances indicate the existence of Atmospheres surrounding the planets Jupiter, Mars and Venus, but are not sufficient to establish the fact with perfect certainty.

VINCE, § 400, &c. BIOT, tom. III. § 37, &c.

SECT.

SECT. X.

OF THE SECONDARY PLANETS.

179. JUPITER, when viewed through a telescope, is found to be accompanied by four small stars, ranged nearly in a straight line, parallel almost to the plane of the ecliptic; and occasionally on the same, or on opposite sides of the planet.

a. Sometimes these small stars pass between us and Jupiter, and their shadows are then seen traversing his disk. Hence it is evident, that both Jupiter and they are opaque bodies, which derive their light from the Sun.

b. Jupiter, accordingly, projects behind him a conical shadow, in which the little stars just mentioned are often immersed; so that they disappear, and are eclipsed, as the Moon is, by falling into the shadow of the Earth.

c. Hence these little stars are to be considered as moons, which circulate round Jupiter; and as they are always observed to move eastward when they are eclipsed, and westward when they pass over the disk, it is evident that their motion is progressive,

progressive, or in the same direction with the motion of the planets round the Sun.

d. The stars which thus circulate round another planet, are called *Satellites*; and also *Secondary Planets*; those that revolve immediately about the Sun, being distinguished by the name of *Primary*.

The Moon is therefore a secondary planet; but has been treated of before the primary planets, because her motions are conspicuous, and very useful for explaining the general phenomena of the heavens. The satellites of Jupiter were discovered by GALILEO, in 1610. *Nuncius Syderius Florentiæ*, 1610.

180. The distances of the different satellites from Jupiter, are nearly in proportion to the angles which these distances subtend from the Earth, when the satellites are at their greatest elongations.

a. The distances are found, on this principle, to be nearly in the ratio of the numbers 6, 9, 15, and 26. The satellites are called *First*, *Second*, &c. in the order of the distances, beginning from that which is nearest to Jupiter.

b. The moments of greatest elongation, and of the beginnings and ends of eclipses, are almost the only times at which the satellites have been observed; and it is thence, that with much ingenuity, and incredible

credible labour, a complete theory of their motions has been established.

181. The time from the middle of one eclipse of a satellite, to the middle of the next, is the time of its synodic revolution, and is equal to the time of its revolution round Jupiter, increased by the time which it takes to describe an arch of its orbit, equal to the arch which Jupiter has described in the same time round the Sun.

a. The synodic revolution of a satellite is subject to variation, as the rate of Jupiter's motion, in his orbit, is not uniform. The greatest equation of Jupiter's orbit is 11° 8′ 2″; and the time of the first satellite's moving over an arch equal to this, is $39^m\ 22^s$; and so much, therefore, may the synodic revolutions of that satellite differ from the mean. Those of the 4th may differ $6^h\ 13^m$ from the same cause. VINCE, *Astron.* vol. I. § 419. LA LANDE, *Astron.* § 2925.

182. The synodical revolutions of the same satellite, are subject to an inequality, depending on the distance of the Earth from Jupiter.

a. When the Earth is nearest to Jupiter, or at the time of Jupiter's opposition to the Sun, if the synodic revolution be *n*, the successive eclipses should happen at the intervals *n*, 2 *n*, 3 *n*, &c.

reckoning

reckoning from the first. But they are found to happen later than those intervals, by quantities which increase progressively, till near the conjunction, when the Earth's distance from Jupiter is the greatest. From this time to the opposition again, a series of anticipations takes place, just equal to the former retardations, and in a reverse order, so as to bring back the eclipses near the opposition, to the time at which they would have happened if none of those inequalities had taken place.

b. The amount of the retardation, from the opposition to the conjunction, is $16^m\ 26^s$ nearly. The same inequality is found in the eclipses of the other satellites. For the first satellite, $n = 42$ hours nearly.

183. The inequality just described, is completely explained, by supposing light to be propagated in succession, and with a velocity by which it passes from the Sun to the Earth in $8^m\ 13^{sec}$.

CASSINI and MARALDI discovered this inequality in the first satellite, and the explanation just mentioned occurred to them, but was rejected, because it implied the existence of the same inequality in the other satellites, which did not appear from their observations to be true. MARALDI also objected to this explanation, that, if just, the inequality must be affected by the eccentricity of Jupiter's orbit. The explanation, however, was revived by ROEMER in 1674, and the inequality was shewn

shewn by HALLEY to belong to all the satellites. It has also been found, that the eccentricity of Jupiter's orbit does affect the inequality in question. MELANDERHJELM, *Astron.* vol. II. p. 584.

184. The same law connects together the periods of the satellites, and their mean distances from Jupiter, which connects the periods of the planets, and their mean distances from the Sun; that is, the squares of the former quantities are as the cubes of the latter.

NEWTONI *Princip. Math.* lib. III. prop. 17.

185. The mean motion of the first satellite, added to twice the mean motion of the third, is equal to three times the mean motion of the second.

a. If m', m'', m''', are the mean motions of the first, second, and third satellite of Jupiter,

$$m' + 2\,m''' = 3\,m''.$$

b. Also if L', L'', L''', are the longitudes of these satellites, $L' - 3\,L'' + 2\,L''' = 180^\circ$. *Mechanique Celeste*, tom. I. p. 342.

186. The orbits of the satellites are not in the same plane with the orbit of Jupiter.

This appears from the duration of the eclipses of the same satellite, when compared with one another.

The

The Table of their inclinations will be afterwards given.

187. The first satellite has been found, from observation, to have an inequality, which, measured by time, is $3^m 30^{sec}$, varying as the sine of the satellite's distance from a certain point in its orbit.

WARGENTIN found out this inequality from observation. VINCE, *Astron.* vol. I. § 421.

188. The second satellite is subject to an inequality of $16^m\ 30^{sec}$ in time, having its changes included in a period of 437 days.

a. The period of 437 days, is also that in which the first three satellites return to the same situation in respect of Jupiter.

The third satellite is subject to an inequality nearly of the same amount, and circumscribed by the same period. This is also from the observations of WARGENTIN.

189. The orbit of the fourth satellite is sensibly elliptical; the equation of the centre is $50' 20''$ nearly, and the place of the higher apsis in 1700 was $10^s 29° 22''$; the longitude of the satellite being $7^s 17° 18' 2''$.

a. From

a. From analogy we may conclude, with considerable probability, that the orbits of all the satellites are either circles or ellipses.

b. Dr HERSCHEL has observed, that the same satellite is more luminous at one time than another, and that the period of these changes is for each satellite the same with the time of its revolution about Jupiter; hence he has inferred, that the satellite also revolves on its axis, in the same time that it revolves about Jupiter.

This is also the law of the Moon's rotation, § 132.

190. The beginning or end of an eclipse of a satellite, marks the same instant of absolute time to all the inhabitants of the Earth, and may therefore be employed for finding the longitude, in the same way with an eclipse of the Moon, § 67, and 145.

a. The immersion of a satellite into the shadow of Jupiter, and its emersion from it, are instants more precisely defined than the beginning or end of a lunar eclipse; and therefore the longitude is more accurately found by the former. The comparison may either be made between the observation and the Tables; or between it and another observation made under a known meridian.

b. The first satellite is the most proper to be observed, its motions being best known, and its eclipses recurring most frequently.

c. Astronomers,

c. Astronomers, who would compare their observations of these eclipses, should be furnished with telescopes very much alike in distinctness and magnifying power. The greater perfection of the telescope will give the immersions later, and the emersions earlier. On the whole, it is best to employ, as DE LAMBRE advises, telescopes of moderate power. *Abregé d'Astron. leçon* 22. § 14.

d. The apparent instant of immersion or emersion, will also be affected by the proximity of Jupiter to the Earth, and by the altitude above the horizon. M. BAILLY has given rules for correcting the effects arising from all these circumstances. See *Theorie des Satellites, par* M. BAILLY, Also *Mém. Acad. Royale des Sciences*, 1771; and *Phil. Trans.* vol. LXIII. LA LANDE, § 3047. VINCE, vol. I. § 460.

191. By observing the time required by the satellites to enter into the shadow, the angles which their diameters subtend at the centre of Jupiter may be computed.

According to WHISTON, the first satellite employs $1^m\ 10^s$; the second, $2^m\ 20^s$; the third, $3^m\ 40^s$; and the fourth, $5^m\ 30^s$, to enter completely into the shadow, when they enter it perpendicularly. LA LANDE, § 3038. From more accurate observations, M. BAILLY computed their diameters, as seen from Jupiter, the first, 60′ 20″; the second, 29′ 42″;

29′ 42″; the third, 22′ 28″; and the fourth, 9′ 39″.

192. The planets Saturn and Uranus have also satellites; the former seven, and the latter six; which, with the Moon, make eighteen secondary planets in the solar system.

a. The fourth of the satellites of Saturn was the first discovered; it was seen by HUYGENS in 1655. The first, second, third, and fifth, were discovered by CASSINI, between the years 1671 and 1684. Other two satellites, which we improperly call the sixth and seventh, though they are nearer to Saturn than any of the rest, were discovered by Dr HERSCHEL in 1789.

193. The connection between the periodic times and the mean distances of the satellites of Saturn, is the same as in the satellites of Jupiter, and in the primary planets.

a. The fifth satellite disappears regularly for about one-half of the time of its revolution round Saturn; and hence NEWTON concluded, that its revolution on its axis is of the same duration with its revolution round Saturn. *Princip. Math.* lib. III. prop. 17.

194. The

194. The six satellites of Uranus have their orbits nearly in one plane, at right angles to the plane of the orbit of the planet.

a. These satellites were discovered by Dr HERSCHEL, in 1807 and 1808.

b. The plane in which these satellites move, passes twice in the year through the Sun; and at those times there may be eclipses, but they cannot be seen unless Uranus be near his opposition.

Eclipses of the satellites of Uranus were visible in 1799, and they will be visible again in 1818.

Ring of Saturn.

195. Saturn is surrounded by a circular ring, concentric with himself, which, being seen obliquely, is of an oval or elliptic form, and shines with the same light as the planet.

a. If we suppose the radius of Saturn, at his mean distance, to subtend an angle of 9″, the interior diameter of the ring, supposing it circular, will be 15″, and the exterior 21″, leaving a void of 6″ between the planet and the ring, and the same quantity for the breath of the ring. Stars have been observed

observed between the ring and the planets. SMITH's *Optics*. DE LA LANDE, tom. III. § 3353.

b. The plane of the ring is in the plane of the equator of Saturn; it is inclined to the orbit of the planet, nearly at an angle of 30°, and remains always parallel to itself.

196. When Saturn is in the longitude of 5^s 20°, or 11^s 20°, the plane of the ring passes through the Sun, and the light then falling upon it edgewise, it is no longer visible to us.

a. This disappearance of the ring has been often observed. LA LANDE, § 3354.

The disappearance of the ring from this cause, lasts only a few days; for when Saturn has passed the node of the ring three or four minutes, or the Sun has risen above its plane by that quantity, the ring becomes visible.

197. The ring also disappears, when its plane passes through the Earth; for its edge, or its thickness, being then directed to the eye, and being too fine to be seen, the planet appears quite round.

a. In this case, the Earth requires to be elevated above the plane of the ring, at least half a degree, before the ring is seen, which makes it continue invisible

seven

seven or eight days before and after the passage of the Earth through its plane.

b. This is what happens with ordinary telescopes; but with the forty foot reflector of Dr HERSCHEL, the ring does not cease to be visible. The Doctor saw the satellites along the margin of the ring, like bright beads threaded on a string.

198. There is a third cause of the disappearance of the ring, viz. the Earth being placed on the side of the ring that is turned from the Sun.

The ring may on these accounts disappear twice in the same year.

LA LANDE, § 3357. *Opuscula* of BOSCOVICH, tom. v. Also SEJOUR, *Anneau de Saturne.*

199. The ring is divided into two by a dark line going all round concentric with the outer and inner circles.

a. This was first observed by CASSINI, *Mém. Acad. des Sciences*, 1715. SHORT, with his twelve foot reflector, observed the same; as Dr HERSCHEL has also done.

b. To Dr HERSCHEL, and some other observers, it has appeared that there are more of these lines than one.

It

It appears, therefore, that there are really two concentric rings; or perhaps more.

200. The ring revolves on an axis at right angles to its own plane nearly, in the same time with the planet itself, or in a little more than ten hours.

This observation we also owe to Dr HERSCHEL.

a. The time of the revolution of the ring is .437 of a day, or $10^h\ 29^m\ 17^s$.

b. It is remarkable, that if a satellite at the mean distance of the middle of the ring, revolved round Saturn, and obeyed the law of KEPLER in respect of the other satellites, it would revolve exactly in $10^h\ 29^m\ 17^s$. BIOT, *Astron. Phys.* tom. III. p. 96.

SECT.

SECT. XI.

OF COMETS.

201. A COMET is a luminous body, which appears in the heavens only for a limited time, seldom exceeding a few months; during which, beside the diurnal motion, of which it partakes in common with the other heavenly bodies, it has always a motion peculiar to itself, by which it changes its place among the fixed stars. Its appearance is usually that of a collection of vapour, in the centre of which is a nucleus, for the most part, but indistinctly defined.

In some Comets, the peculiar motion is progressive, in others retrograde. In the same comet, the motion is all nearly in one plane; but in different Comets, these planes make all different angles with the ecliptic. At the beginning and end of the appearance of a Comet, it deviates from the plane in which the middle part of its course lies. NEWTON *de Systemate Mundi*, § 59. The Comets have no parallax, and are therefore certainly beyond the limits of our atmosphere.

202. A Comet, when it first appears, is usually surrounded by a faintly luminous vapour, to which the name of *Coma* has been given. As the Comet approaches the Sun, the coma becomes more bright, and at length shoots out into a long train of luminous transparent vapour, very much resembling a *streamer*, and in a direction opposite to the Sun. This forms the *Tail* of the Comet.

As the Comet retires from the Sun, the tail grows less, and resumes nearly its first appearance. Those Comets which never come very near the Sun, have nothing but a coma or nebulosity round them during the whole time of their continuance.

The tail is always transparent, so that the stars are distinctly seen through it, as they are even said to have been in some instances through the central part, or what was supposed the nucleus of the Comet.

The length and form of the tail are very various. Sometimes it is only a few degrees, at others it is more than a quadrant. In the great Comet which appeared in 1680, the tail subtended an angle of 70° ; in that of 1618, an angle of 104°.

The tail sometimes consists of diverging streams of light; that of 1744 consisted of six, all proceeding from the head, and all a little bent in the same direction.

The

The Comet of 1811 was remarkable for its beauty. The tail was composed of two diverging beams of faint light, slightly coloured, which made an angle from 15° to 20°, and sometimes much more, and were bent outward. The space between was comparatively obscure. *Abrégé d'Ast. par* DE LAMBRE, *leçon* 21. § 57.

203. A Comet remains so short a time in sight, and describes so small a part of its course within our view, that, from observation alone, without the assistance of hypothesis, we should not be able to determine the nature of its path. The hypothesis most conformable to analogy is, that the Comet moves in an ellipsis round the Sun placed in one of the foci, and that the radius vector from the Sun to the Comet describes areas proportional to the times.

As the ellipse in which a Comet moves is evidently very eccentric, it will coincide very nearly with a parabola, at its vertex, or for all the time that a Comet remains in sight.

204. If it be supposed that the Comet describes an ellipsis or a parabola, in conformity to the laws of KEPLER, then from three geocentric places, known by observation, the orbit may be determined.

The first solution of this problem was given by NEWTON, *Princip.* lib. III. prop. 41. He calls it *Proble-*

ma

ma longè difficillimum, and it may, therefore, be readily believed, that the solution does not fall within the limits of an elementary treatise. At the same time, a very simple geometrical problem is the foundation of it. *Arith. Universalis.* The determination of the orbit implies that of the five quantities, which are its elements :

1. The inclination of the orbit. 2. The position of the line of the nodes. 3. The longitude of the perihelion. 4. The perihelion distance from the Sun. 5. The time when the Comet is in the perihelion.

NEWTON'S solution being a laborious and indirect approximation, the problem has been attempted by many others. LA CAILLE, *Astron.* § 775, &c. BOSCOVICH, *Opera*, tom. III. p. 14. &c. The solutions of BOSCOVICH and LA PLACE are illustrated by Sir H. ENGLEFIELD, in his *Determination of the Orbits of Comets*, Lond. 1793.

LA LANDE has given a mechanical construction, that serves for finding the orbit nearly, § 3127. ; it is followed by VINCE, § 653. LAMBERT has demonstrated some remarkable properties of the orbits of Comets, in a work entitled *Insigniores Orbitæ Cometarum proprietates*, Aug. Vind. 1761, 8vo. See also LA GRANGE, *Mém. de Berlin*, 1783.

The most perfect solution of all is supposed to be that of LA PLACE, *Mechanique Celeste.*

The

The latest, and in practice one of the best, is that of DE LAMBRE, *Abrégé d'Astron. leçon* 21. § 24.

205. The only Comet which is known with absolute certainty to have returned, is that of 1682, which, conformably to the prediction of Dr HALLEY, appeared in 1759.

Dr HALLEY was led to this prediction by observing, that a Comet had appeared in 1607, and another in 1531, and that the elements of their orbits, when calculated from the observations made on them, agreed nearly with those of the Comet of 1682, the period being between seventy-five and seventy-six years.

Though there can be little doubt that it was the same Comet which was seen at all the four dates just mentioned, the appearances were considerably different. In 1531 the Comet was of a bright gold colour; in 1607 it was dark and livid; it was bright again in 1682, and obscure in 1759. PINGRE', *Cometographie*, tom. II. p. 189.

The return of some of the other Comets is probable, though not certain.

The great Comet of 1680, was supposed by Dr HALLEY to have a period of 575 years, and to be the same which had appeared a little before the death of JULIUS CÆSAR, in the year 44 A. C.; again, in the reign of JUSTINIAN, in the year 531 P. C.; and

and in 1106, in the reign of HENRY I. At all these periods, appearances of a great and terrible Comet are recorded, but no such observations was made as can ascertain their identity completely. *Synopsis Astronomiæ Cometicæ*, subjoined to HALLEY'S *Astronomical Tables*.

206. The Comet of 1680, mentioned above, is remarkable for having approached nearer to the Sun than any other that is known. At its perihelion, its distance from the Sun was only $\frac{1}{166}$th part of the Earth's. It descended to the Sun with great velocity, and almost perpendicularly, and ascended in the same manner, remaining in sight for four months.

When this Comet was in the perihelion, the diameter of the Sun must have subtended an angle of more than 100 degrees. See many interesting particulars with respect to it, *Princip.* tom. III. *prop.* 41. *at the end.*

The phenomena of the tails of Comets, shew the celestial spaces to be void of resistance.

Some Comets have come very near the Earth. A Comet in 1472 is said by REGIOMONTANUS to have moved over an arch of 120 degrees in one day; and another in 1760 moved over an arch of 41 degrees in the same space of time. As neither of them could probably have described in its orbit more

more than an arch of a few degrees, these extraordinary apparent changes of place can only have proceeded from vicinity to the Earth.

207. It appears that Comets contain very little matter, and have but a very feeble action on other bodies.

In the year 1454, a Comet is said to have eclipsed the Moon; so that it must have been very near to the Earth; yet it had no sensible effects. The Comets just mentioned produced none. A Comet in 1770 came very near to the satellites of Jupiter, but caused no derangement in the system.

208. The number of Comets observed and recorded, with more or less accuracy, exceeds 350.

PINGRE', who wrote in 1783, enumerates 321; and 32 are now to be added.

Of these, not so many as a third have been observed with such accuracy, as to allow the elements of their orbits to be ascertained. The elements of 98 have been computed, going back as far as the year 837, and coming down to the Comet of 1807. See FERGUSSON'S *Astronomy*, (BREWSTER'S edit.) vol. II. p. 360.

SECT.

SECT. XII.

OF THE ABERRATION OF LIGHT, AND THE NUTATION OF THE EARTH'S AXIS.

Aberration of Light.

209. If a ray of Light, coming in a straight line, and with a given velocity, pass through a tube, also moving in a straight line, with a given velocity, and remaining parallel to itself; the path of the ray, relatively to the tube, is the diagonal of a parallelogram, the sides of which are proportional to the velocities of the Light, and of the tube, and in the same straight lines, one of the velocities only, being estimated in a direction opposite to its own.

This is evident from the composition of motion.

We may conceive what is here supposed to be actually the case, with a ray of light coming from a fixed star, and passing through a telescope, or a tube, furnished with plain sights; the tube having the same motion with the earth in its orbit, which, for a short time, may be regarded as rectilineal and uniform. The path of the ray in the tube, is therefore

therefore not directed to the point from which the ray comes, and the line of *collimation* (the line in the tube along which the object is viewed) must be inclined, so as to have the direction of the path of the ray; or of the diagonal just mentioned.

The angle thus contained between the line drawn to the object, and the line of the tube in which the object is seen, is called the *Aberration*.

210. When the motion of the tube, or of the earth, is perpendicular to the motion of the ray, the aberration is a *maximum*, and is measured by the arch which the earth describes in its orbit, in the time which light takes to come to it from the sun.

In 8′ 13″, light moves over the radius of the earth's orbit: and, in the same time, the earth moves over an arch of 20″.232, which is therefore the *maximum* of the aberration; and this is found by observation to be accurately true.

211. Every star appears to describe an ellipsis in the heavens, of which the true place of the star is the centre; the semi-transverse axis, 20″.232, in the direction of a tangent to the parallel of latitude of the star; and the semi-conjugate axis, 20″.232 × sin Lat. of the star.

SIMPSON'S *Essays*, 1740, p. 1. &c.

If

If the star is in the pole of the ecliptic, it describes a circle, with the radius 20″.232, having the pole for its centre.

If the star is in the ecliptic, it describes a straight line in that plane, and extending 20″.232 on each side of the star.

The purposes of practical astronomy require, that the change made by the aberration on the longitude and latitude, and on the right ascension and declination of a star, should be computed.

212. If L be the longitude of the sun at any time, and L′ the longitude of a star, the aberration of the star in longitude is

$$\frac{-20''.232 \times \cos(L'-L)}{\cos \text{Lat.}};$$

and the aberration in latitude is

$$20''.232 \times \sin(L'-L)\sin \text{Lat.}$$

These formulas were first given by Clairaut. *Mém. Acad. de Scien.* 1737.

See also Cagnoli, *Trig.* § 1529. La Lande, § 2823. and De Lambre, *Ast. leçon* 19. § 20, 21. &c.

213. If A be the right ascension, and D the declination of a star, L being the sun's longitude,

tude, as before, the aberration of the star in right ascension is

$$-\frac{19''.17 \times \cos(A-L) - 0''.83 \times \cos(A+L)}{\cos D};$$

and the aberration in declination is,

$$\sin D(19''.17 \sin(A-L) - 0''.83 \sin(A+L) - 8'' \cos L \times \cos D).$$

CAGNOLI, *ibid.*

DE LAMBRE, *ibid.*

From these four formulas, all the effects of aberration may be computed. Tables have been constructed for facilitating the calculations. VINCE, *Ast.* vol. I. § 513. & 522.

The discovery of the aberration was made by Dr BRADLEY, in 1725, from observation alone. He has given an account of it, in a letter to Dr HALLEY, *Phil. Trans.* N° 406. A very full extract of this letter is in VINCE's *Astron.* vol. I. § 503. After observation had satisfied him of the changes above stated, in the latitudes of different stars; he discovered the cause to be the motion of light, combined with the motion of the earth, as above explained.

214. From the phenomena of the aberration, the motion of the earth in its orbit is more directly proved than from any other fact in astronomy.

Though

Though it is proved to demonstration, from facts above enumerated, that the Earth is far from being the centre of the planetary motions, yet all the appearances hitherto mentioned, are consistent with what is called the TYCHONIC *System* of the heavens, (from its inventor TYCHO BRAHE'), in which the Sun, accompanied by the planets, revolves in an orbit round the Earth. This system, which its want of simplicity renders suspected, is entirely overturned by the fact of the aberration, and the motion of the Earth completely established.

When the aberration was first discovered, it was thought that the velocity of light, as inferred from the eclipses of the satellites of Jupiter, did not perfectly agree with it. It has, however, been found, from more accurate comparisons, that they perfectly coincide.

215. It appears, that the light of the heavenly bodies traverses the spaces between them and the earth with the same uniform velocity.

The reflected light from the satellites, travels with the same velocity with the direct light of the fixed stars; and the velocity of this last is the same from whatever distance it comes.

There is reason to think, that light is accelerated by the action of transparent bodies. If so, these bodies also impart to light a velocity in the direction of the earth's motion, and proportional to the increase

increase of velocity which they produce, because their action does not change the quantity of the aberration.

216. The aberration considered above, is that of a fixed star, or of an immoveable body. In the planets, there is another source of aberration, in the motion of the planet itself, equal to the variation of its geocentric place, in the time that light takes to move from the planet to the earth.

The simplest way of making allowance for this variation, is to compute the place of the planet for an instant preceding the given instant, by the time that light takes to move from the planet to the earth. This was the method used by Dr MASKELYNE. VINCE, *Ast.* vol. I. § 527.

No attention has been paid here to the velocity which places on the Earth's surface derive from the motion of the Earth on its axis. This is too small to produce a sensible effect, as is shewn by DE LAMBRE, *Astron. leçon* 19. § 23.

Nutation of the Earth's Axis.

217. A small inequality, which has been observed in the precession of the equinoxes, and in the mean obliquity of the ecliptic, is known by the name of the *Nutation.*

This

This inequality, which affects all the heavenly bodies, was discovered by Dr BRADLEY, while employed in verifying his theory of the aberration.

The period of the changes of this inequality was observed to be eighteen years nearly, the same with the period of the revolution of the Moon's nodes; and it was soon found, that the quantity of the inequality depended on the place of the node. Certain theoretical considerations thus led to the discovery of the precise form which this inequality assumes, when analytically expressed, and reduced into a formula. It is not certain that observation alone would have led to this conclusion; but there is no doubt that the places of the heavenly bodies, when corrected by means of it, agree much better with observation than when the correction is not applied.

218. The phenomena of the nutation may be represented by supposing, that while a point, which may be considered as defining the mean place of the pole of the equator, describes a circle in the heavens, round the pole of the ecliptic, at a distance from it equal to the mean obliquity of the ecliptic, and with a retrograde motion of 50″ annually; another point, representing the actual pole of the equator, moves round the former at the distance of 9″, so as to be always 90° more easterly than the Moon's ascending

ing node. The inequality thus produced in the precession of the equinoxes, and in the obliquity of the ecliptic, will exactly agree with the appearances as observed.

It was by this construction, that Dr BRADLEY represented the irregularities he had observed.

From the oscillatory motion thus ascribed to the pole of the equator, or to the axis of the Earth, the name of *nutation* is derived.

219. If N be the longitude of the moon's ascending node, $+ 9''.6 \cos N$ is the variation in the obliquity of the ecliptic produced by the nutation, and $- 17''.946 \sin N$ is the variation in the precession, or in the equinoctial points.

The last of these equations affects the longitude of all the heavenly bodies equally.

DE LAMBRE, *Abrégé d'Ast. leçon* 20. § 16.

220. If A be the right ascension of a star, and D its declination, the nutation in right ascension is

$$\tan\big(D - 8''.373 \cos(A - N) - 1''.227 \cos(A + N)\big);$$

and the nutation in declination is

$$+ 8''.373 \sin(A - N) + 1''.227 \sin(A + N).$$

The

The effect on the declination, therefore, does not involve the declination itself, and is the same for all stars having the same right ascension.

221. The inequalities here considered, do not affect the position of the stars relatively to one another, nor to the plane or pole of the ecliptic; they affect their position only relatively to the plane of the equator, or to the position of the Earth's axis.

It was from this general fact, combined with the relation observed between these inequalities, and the motion of the moon's nodes, that Bradley was led to the construction given above. In separating between the inequalities belonging to the nutation, and those depending on the aberration, (which observation always presented as combined together), he has displayed great acuteness and sagacity.

See Woodhouse, *Astron.* chap. 17.

Sect.

SECT. XIII.

DIMENSIONS OF THE SOLAR SYSTEM.

HITHERTO, the distance of the Sun from the Earth has served as the unit, by which we have measured all other distances in the planetary system. It now remains, (in order to have a precise idea of those distances), to compare this unit with the diameter of the Earth, and of consequence with the known measures in which that diameter has already been expressed. This depends on the parallax of the Sun, which has been shewn to be less than 10″, and, on account of its smallness, difficult to be ascertained. The method which first presents itself, does not lead to any thing more precise than the limit just mentioned.

222. Since the ratios of the distances of the planets from the Sun, to the distance of the Earth from the Sun, are known, if the parallax of any of the planets were discovered, that of the Sun would, of consequence, become known.

This follows, readily, from the laws of the planetary motions.

Mars, when in opposition to the Sun, is nearer the Earth than any other of the superior planets, and his parallax of consequence is the greatest. From the opposition of October 1751, above referred to, and observed by LA CAILLE, at the Cape, the parallax of Mars was determined to be 24″.6; and his distance from the Earth 3371 semidiameters of the latter. But, from the place of Mars in his orbit, his distance from the Earth at that time was .4354 of the parts of which the Sun's distance from the Earth is 1; so that the parallax of the Sun is equal to that of Mars, multiplied by the above decimal, and is therefore 10″.69; and therefore the distance of the Sun is 19226 semidiameters of the Earth.

This, however, cannot, any more than the former determination, be considered in any other light than as a limit, which the Sun's distance probably exceeds.

223. The planet Venus, at the inferior conjunction, approaches nearer to the Earth than Mars does in opposition; and, therefore, if Venus can then be observed, which she may be when she passes over the disk of the Sun, that observation will afford the best means of ascertaining the Sun's parallax.

This

This was first remarked by Dr HALLEY, and since that time two transits of Venus over the Sun have happened, which astronomers have taken the greatest pains to observe with accuracy. The general principle which connects this method of finding the parallax with the more elementary methods already explained, is all that can be given here. The details of the calculations must be left to the Treatises and *Mémoires* which treat of them particularly.

224. Let S be the Sun (fig. 21.) and E the Earth, both supposed at rest, while Venus at V moves westward in her orbit, with the sum of her own angular velocity and that of the Sun. Let O and O′ be the stations of two observers on the surface of the Earth, who see the transit begin when Venus is at the points V and V′ of her orbit. If the difference of the longitude of the two observers be known, the time that Venus has taken to move over the arch VV′ is also known, and therefore the arch VV′ is given, or the angle which the line OO′ subtends at the distance of the Sun. But OO′ itself is given, and its ratio to the radius of the Earth; therefore the angle which that radius subtends at the Sun, or the horizontal parallax of the Sun, is given.

This

This construction supposes the observers O and O′ to be either exactly, or nearly in the plane of the orbit of *Venus*, but it may be extended to cases in which that condition does not take place. It requires, too, that the longitude of the places of observation should be accurately known. To avoid the necessity of this determination, the durations of the transit, as seen from different stations, have been preferred, for ascertaining the parallax. If we suppose observers, situated in respect of one another, so that the line which Venus is seen to describe on the Sun's disk, is longer at the one station than the other; the duration of the transit will be proportionally greater, and the difference will evidently depend on the distance of the observers from one another, estimated in the direction perpendicular to the lines which Venus traces out on the surface of the Sun. The differences of duration, therefore, depend on the parallax of the Sun, or on a *function* of it; and therefore when that function is known, the parallax may be inferred, from the comparison of one value of it with another. See WOODHOUSE, *Astron.* p. 378, &c.

The transit of 1769 was observed at Wardhus or the North Cape, and also at Otaheité in the South Sea, and was found to be longer at the former than at the latter by 23′.10″. This difference, supposing the parallax to have been 8″.83, should have amounted to 23′.26″.95; and hence the parallax is deduced = 8″.72. VINCE, *Ast.* I. § 622. Dr MAS-

KELYNE'S

KELYNE's method of calculating the parallax from the duration of the transit, is there given.

See also on this subject, LA LANDE, *Astron.* vol. II. liv. XI.

225. From a mean of the observations of the transit of 1769, taking it as calculated by SHORT, EULER, &c. the parallax of the Sun, for his mean distance, has been found to be 8″.73; and the mean distance, therefore, = 23659 semidiameters of the Earth, or 93595000 miles.

VINCE, *Astron.* vol. I. § 629.

Thus the scale is determined according to which the distances of the planets have been already set down, § 173.

SECT.

SECT. XIV.

OF THE ANNUAL PARALLAX AND DISTANCE OF THE FIXED STARS.

226. The fixed stars, as has been already shewn, have no parallax with respect to this Earth, or any line that can be measured on its surface; and their distance is so great, that it is yet doubtful whether they have any parallax, even with respect to the orbit of the Earth round the Sun.

A fixed star not only occupies exactly the same place in the heavens, from whatever point of the Earth's surface it is observed, but it does so within a quantity so small as to be hardly measurable, even when viewed from opposite extremities of a diameter of the Earth's orbit.

On the supposition that the star does change its situation, when so viewed, the angle which measures that change is called the *Annual Parallax* of the Star.

227. If

227. If a fixed star had an annual parallax that was sensible, it would appear to describe an ellipsis, of which the greater semi-axis was equal to that parallax, and the semi-conjugate equal to the same, multiplied into the sine of the latitude of the star.

The centre of this ellipse would be the place of the star, seen from the Sun; the conjugate would have the direction of a circle of latitude, passing through the star and the pole of the ecliptic.

228. If L be the longitude, and λ the latitude of a star, of which the annual parallax, when a *maximum*, is p; then, if the longitude of the Sun is S, the parallax of the star, in latitude, will be

$$-p \sin \lambda \cos (L-S);$$

and its parallax in longitude

$$\frac{p}{\cos \lambda} \sin (L-S).$$

De Lambre, *Abrégé, leçon* 19. § 29.

The effects of aberration and of parallax, are both to make the fixed stars describe ellipses, but quite different, in magnitude and position. Boscovich, however, has demonstrated, that a fixed star, under the influence of both these causes, will still appear to describe an ellipsis about its true place.

Dissertatio

Dissertatio de annuis fixarum aberrationibus. Romæ, 1742. LA LANDE, *Astron.* § 2857.

229. When two stars appear very near to one another, or when their distance subtends a very small angle, the variations in that angle, at opposite seasons of the year, may serve to determine the parallax of the fixed stars.

This method of ascertaining the distance of the fixed stars, seems to have been first thought of by GALILEO, *Syst. Cos.* Dial. 3. Dr HERSCHEL has also recommended it; and the double stars which he has discovered, as well as the *Lamp Micrometer* which he has invented, give hopes that it may prove successful.

230. From the consideration of the quantity of the light of the fixed stars, compared with the quantity of the light of the Sun, it has been concluded, that the parallax of a star of the second magnitude, is not more than $\frac{1}{3}$ of a second; and of a star of the sixth magnitude, not more than $\frac{1}{20}$ or $\frac{1}{30}$ of that quantity.

The very ingenious paper in which these conclusions are deduced, is by the Reverend Mr MITCHELL. *Phil. Trans.* vol. lvii. p. 234, &c.

231. The

231. The attempts of astronomers to discover the annual parallax of the fixed stars by direct observation, have not yet been perfectly successful; and it is therefore probable, that the quantity sought for does not exceed one second.

Dr BRINKLEY is of opinion, that he has ascertained the annual parallax of *α Lyræ* to be 2″.52. *Phil. Trans.* 1810, p. 204.

The attempts to discover the parallax, have chiefly been made on smaller stars, which, being probably more distant than *α Lyræ*, their parallax has been too small for observation. The observation of Dr BRINKLEY is of great authority, but, while single, cannot be considered as perfectly decisive.

If we suppose the annual parallax not to exceed 1″, the distance of the fixed stars cannot be less than 206265 times the radius of the Earth's orbit. As light traverses the latter in 8′.13″, it will require 3 years and 79 days to come from a fixed star to the Earth.

This may be supposed true for stars of the second magnitude, even if those of the first have a parallax of two seconds.

Though this distance is immense, it is probably small compared to that of the most remote of the bodies which

which we see in the heavens. As it cannot be doubted, that the fixed stars are luminous bodies like the Sun, it is probable that they are not nearer to one another than the Sun is to the nearest of them. When, therefore, two stars appear like a double star, or very near to one another, the one must be placed far behind the other, but nearly in the same straight line, when seen from the Earth. The same must hold, at least in a certain degree, wherever a great number of stars are seen concentrated in a small spot. In the starry nebulæ, therefore, such as the Milky Way, which derive their light from the number of small stars, appearing as if in contact with one another, it is plain, that the most distant of these must be many thousand times farther off than the nearest, and light must, of course, require many thousand years to come from them to the Earth. The poet, perhaps, has been taxed with exaggeration, who spoke of

> " Fields of radiance, whose unfading light
> " Has travelled the profound six thousand years,
> " Nor yet arrived in sight of mortal things."

Yet the fields which he describes, are far within the circle to which the observations of the astronomer extend !

AP-

APPENDIX.

ON THE METHOD OF DETERMINING BY OBSERVATION, THE CONSTANT CO-EFFICIENTS IN AN ASSUMED OR GIVEN FUNCTION OF A VARIABLE QUANTITY.

Observation makes known the places of the heavenly bodies only for instants, separated from one another by certain intervals of time; but the purposes of science require, that they should be determined for any time whatever, or for every instant. This is rendered practicable by the consideration, that the magnitude which determines the place of any body, its longitude or latitude for example, is a variable quantity, between which and the time (a variable quantity also) a certain relation continues always the same. This relation may therefore be expressed by an equation, which, for every value of one of the variable quantities, will give the corresponding value of the other; so that the one of them may be assigned in terms of the other, or in what is called a *Function* of it. The form of this function is sometimes known from theory or analogy, and at other times it is wholly unknown. In the

the latter case, we must assume the simplest function that can represent the observations, and this naturally consists of a series of terms, proceeding according to the powers of one of the variable quantities, with co-efficients which are constant, but unknown quantities, to be found from the observations. This is called the method of *Interpolation*, because it inserts a term in the midst of a number of others.

232. If x and y are two variable quantities, of which several values have been determined from observation; if y be assumed equal to a series of the powers of x, beginning from 0, and going on to as many terms as there are observations, viz. $y = A + Bx + Cx^2 + Dx^3$, &c.; then, if for y and x, be put their corresponding values, as determined by observation, as many equations will arise as there are unknown co-efficients, A, B, C and D to be found, from which they will become known.

The most useful interpolations are, when the time is one of the unknown quantities, and when the intervals between the observations are equal, as is supposed in what follows:

233. Let α, $\acute{\alpha}$, α'', α''', α'''', be any number of quantities determined from observations made at

at the times 0, m, $2m$, $3m$, $4m$, &c. and let the differences of α, α', α'', &c. be taken in their order, and the difference of these differences, as on the margin, so that β is the first of the first differences, γ of the second, δ of the third, and ε of the fourth, &c.

$$\begin{array}{l} \alpha, \alpha', \alpha'', \alpha''', \alpha'''', \\ \beta, \beta', \beta'', \beta''', \\ \gamma, \gamma', \gamma'', \\ \delta, \delta', \\ \varepsilon \end{array}$$

$$y = \alpha - \frac{\beta}{m} x + \frac{\gamma}{2m} x \left(\frac{x}{m} - 1\right)$$

$$- \frac{\delta}{6m} x \left(\frac{x}{m} - 1\right)\left(\frac{x}{m} - 2\right)$$

$$+ \frac{\varepsilon}{24m} x \left(\frac{x}{m} - 1\right)\left(\frac{x}{m} - 2\right)\left(\frac{x}{m} - 3\right),$$

&c.

where the law of continuation is evident.

This is Sir ISAAC NEWTON's theorem, expressed algebraically. *Math. Princip.* lib. III. lemma 5.

a. If there are only two observations, so that the series of differences goes no farther than β, then

$y = \alpha - \beta \cdot \frac{x}{m} = \alpha - \frac{\alpha - \alpha'}{m} x$; and this is no other than the common rule for finding a proportional part.

b. If

b. If there are three observations, $y = \alpha - \beta \cdot \frac{x}{m} + \frac{\gamma}{2} \cdot \frac{x}{m}\left(\frac{x}{m} - 1\right) =$

$$\alpha - \frac{2\beta + \gamma}{2m} x + \frac{\gamma}{2m^2} x^2.$$

c. Attention must be paid to the signs of the differences. The succeeding quantity is here always subtracted from the preceding, and therefore if the former is the greater, the difference will be negative. On the subject of interpolation, See NEWTON, as above quoted. STIRLING on *Infinite Series*; LA CAILLE, *Astron.* p. 69.; *Nautical Almanack*, explanation at the end.

Interpolations of this kind, cannot be supposed to extend far beyond the interval within which the observations are contained, unless it shall appear, on continuing the observations, that the formula applies to them, without requiring the addition of new terms. As long as the addition of new observations requires an alteration in the formula, it is certain that the true law is not discovered.

234. Let it now be supposed, that the form of the function is known from theory, but that the constant quantities which enter into that function, as co-efficients of the different terms, are to be found from observation; required, considering that every observation is liable to error, in what way these quantities may be most accurately determined.

There

There are two cases of this problem, one, when the constant quantities are yet entirely unknown, another, when they are already approximated, and only require to be farther corrected. In the first case, the process is very simple.

235. Substitute the quantities known by observation for y and x, in the given formula, (each observation being supposed to afford a value both of x and of y), and thus, as many equations will be obtained, as there are observations. Then let there be composed, from the addition of these into separate sums, as many equations as there are quantities to be found. The solution of these equations will give the quantities required, and because, in a multitude of observations, the errors must in a great measure compensate for one another, the results thus obtained will be more accurate, *cæteris paribus*, the greater the number of the equations combined.

The postulate on which this rule proceeds is, that though each of the given equations is incorrect, on account of the unavoidable errors of observation, there is nothing that determines the amount of the errors to be on one side more than another, or in excess rather than defect. By adding a number of the equations together, the errors may therefore be expected, in part at least, to balance

one

one another, so that a more correct equation is obtained than any one of those of which it is composed.

Some combinations of the equations, however, are more advantageous than others. A good rule to be followed in the selection of them, is to unite in one sum all the equations in which the co-efficients of the same unknown quantity, A for instance, are the greatest, ordering it so that all those co-efficients may have the same sign. Do the same with respect to B and C, and the other quantities to be found. Thus, each of the unknown quantities will appear in one of the equations with the greatest co-efficient possible, and will be determined from the cases in which it had the greatest share of the total effect.

Suppose the general formula to be

$$y = \text{A} \sin x + \text{B} \sin 2x,$$

and that from observation we have eight values of x and y, viz.

Values of x.	Values of y.
140°	73′.5
135	80 .2
130	87 .0
125	94 .1
120	99 .5
115	104 .5
110	107 .5
105	110 .2

1 Hence,

Hence,

$$
\begin{aligned}
.6428\,A - .9848\,B &= 73.5 \\
.7071\,A - 1.0000\,B &= 80.2 \\
.7660\,A - .9848\,B &= 87.0 \\
.8191\,A - .9337\,B &= 94.1 \\
.8660\,A - .8660\,B &= 99.5 \\
.9063\,A - .7660\,B &= 104.5 \\
.9397\,A - .6428\,B &= 107.5 \\
.9660\,A - .5000\,B &= 110.2
\end{aligned}
$$

By adding the last four into one, and also the first four, we get

$$3.6780\,A - 2.7748\,B = 421.7, \text{ and}$$
$$2.9350\,A - 3.9033\,B = 334.8;$$

and therefore,

$$A = \frac{3.9033 \times 421.7 - 2.7748 \times 334.8}{3.678 \times 3.9033 - 2.935 \times 2.7748},$$

$$\text{or } A = 1^{\circ}.55'.00.$$

In like manner, $B = 1'.2$; so that the equation becomes,

$$y = (1^{\circ}\ 55')\ \sin x + (1'.2)\ \sin 2x.$$

This is nearly the equation of the centre in the Earth's orbit.

In this way all the elements of any of the planetary orbits may be determined *simultaneously*, or may be corrected if they are already nearly known. In the construction of Astronomical Tables, the number of equations combined has sometimes amounted to several thousands.

 Suppose,

Suppose, for example, that the formula

$$(121^\circ) \sin x - (65^\circ.5) \sin 2x = y,$$

was reduced into a table, and that by a comparison with observation, it was required to correct the quantities 121°, and 65°.5, that is, A and B, according to the former notation.

Suppose also that the observations were made when the values of x were 30°, 45°, 60°, 75°, 90°, 105°, 120°; and that the errors in the Tables were in these instances found to be +.318, —.334, —.083, +.044, +05, —.021, —.084; then, calling a and b the corrections on A and B, we have

$$\begin{aligned}
.500\,a + .866\,b &= +.318 \\
.707\,a + 1.000\,b &= -.334 \\
.866\,a + .866\,b &= -.083 \\
.966\,a + .500\,b &= +.044 \\
1.000\,a + 0 &= +.05 \\
.966\,a - .500\,b &= -.021 \\
.866\,a - .866\,b &= -.084.
\end{aligned}$$

By adding all these equations together,

$$5.971\,a + 1.076\,b = -.299.$$

Then, ordering all the seven original equations, so that the terms involving b shall be affirmative, we obtain

$$2.207\,a + 4.598\,b = +.1.$$

From these, $a = -.0527$, and $b = +.047$.

This

This method of determining the co-efficients of a given function, or correcting them from observation, by means of what are called *Equations of Condition*, is said to have been invented by TOBIAS MAYER of Göttingen, and employed in the construction of his Lunar Tables. He has, however, given no account of it in any of his works; and DE LAMBRE says, that his own Astronomical Tables are the first in which it is certainly known to have been used. *Astronomie*, tom. II. § 85. It is now generally adopted by astronomers, and may be introduced with great advantage into many experimental investigations *. It has also been improved on by LE GENDRE, who determines the quantities sought, so that the sums of the squares of the errors shall be a *minimum*. The investigation is however too difficult to be introduced into an elementary treatise.

OUT-

* Any equation, expressing the relation that obtains among the co-efficients of another equation, is called an *Equation of Condition*; the phrase, however, is usually confined to *differential* equations.

OUTLINES

OF

NATURAL PHILOSOPHY.

ASTRONOMY.

PART II.

PHYSICAL ASTRONOMY.

SECTION I.

OF THE FORCES WHICH RETAIN THE PLANETS IN THEIR ORBITS.

236. If a body gravitating to a fixed centre, have a projectile motion impressed on it, in a line not passing through that centre, it will move in a curve; and the straight line drawn from the body to the centre, will describe areas proportional to the times.

Princip. Math. lib. I. prop. 1.

a. Conversely,

a. Conversely, if a body move in a curve, so that the line drawn from it to a fixed point, describe areas proportional to the times; the body gravitates to that point, or tends continually to descend to it.

b. The velocities of a body in different points of the curve which it describes about a centre of force, are inversely as the perpendiculars drawn from the centre to the tangents at those points.

c. By comparing this proposition with the first of KEPLER's laws, it follows, that the primary planets all gravitate to the Sun, and that the secondary planets gravitate, each to its primary; and thus the laws of Dynamics are immediately extended to the motions of the heavenly bodies.

d. If the curve in which the body moves returns into itself, it is called an *Orbit*, as in the case of the planets; if it does not, or if it is not known that it does it, is called a *Trajectory*.

237. If of two bodies gravitating to the same centre, one descend in a straight line, and the other revolve in a curve; then, if the velocities of these bodies are equal in any one case, where they are equally distant from the centre, they will always be equal where they are equally distant from it.

Princip. lib. I. prop. 40.

a. If,

a. If, instead of descending, the body that has the rectilineal motion should ascend from the centre, while the revolving body retires from it, the same thing holds of their velocities.

b. When in these propositions a body is said to gravitate to a centre, or to be urged toward it by a centripetal force, no supposition is implied concerning the nature of that force, or the cause from which it proceeds. Nothing more is understood, than the mere existence of a fact, like the tendency of bodies to fall at the surface of the Earth.

238. If a body descending from rest at a given point in a straight line, toward another given point to which it gravitates, have its velocity always proportional to the square root of its distance from the first of these points, divided by the square root of its distance from the second, it is urged toward this last point, by a force which is inversely as the square of the distance from it.

a. Let A, fig. 22. be the point from which the body falls, C the centre to which it tends; let AD, perpendicular to AC, be equal to a given line; and as the centripetal force at A to the centripetal force at any point B in the descent of the body, so let AD be to the perpendicular BE, and let DEF be the curve which passes through the extremities of all the perpendiculars so drawn; then twice

twice the area ABED is equal to the square of the velocity at B, (§ 100. vol. I.). Let this velocity $= v$; having bisected AC in G, let the velocity in G be $= c$. Draw the perpendicular $b\,e$ indefinitely near to BE. Then $v^2 : c^2 :: \frac{AB}{BC} : \frac{AG}{GC} = 1$, or $v^2 = c^2 \times \frac{AB}{BC}$. But $v^2 =$

$2\,ABED$, therefore $2\,ABED = c^2 \times \frac{AB}{BC}$, or

since $AB = AC - CB$, $2\,ABED = c^2 \times \frac{AC - CB}{BC}$

$= c^2 \left(\frac{AC}{BC} - 1\right)$.

For the same reason, $2\,A\,b\,e\,D = c^2 \times \left(\frac{AC}{b\,C} - 1\right)$,

and therefore the area $2\,B\,b\,e\,E = c^2 \left(\frac{AC}{b\,C} - \frac{AC}{BC}\right)$

$c^2 \left(\frac{AC \times B\,b}{BC^2}\right)$. Now, $2\,B\,b\,e\,E = 2\,B\,b \times BE$,

wherefore, $2\,B\,b \times BE = c^2 \left(\frac{AC \times B\,b}{BC^2}\right)$, and

dividing by $B\,b$, $2\,BE = c^2 \times \frac{AC}{BC^2} = c^2 \times \frac{2\,AG}{BC^2}$;

therefore, $BE = c^2 \times \frac{AG}{BC^2}$. But BE represents

the centripetal force at B any point in the line AC, and both c and AG are given, therefore the centripetal force at B is inversely as the square of BC, the distance from the centre of force.

Therefore,

Therefore, if the velocity of a body attracted to a centre, be as the square root of the distance it has fallen through, divided by the square root of the distance between it and the centre, the centripetal force by which it is impelled is inversely as the square of its distance from that centre.

Hence also, if the centripetal force at G be = F, and if $CG = a$, $F = c^2 \times \frac{AG}{AG^2} = \frac{c^2}{a}$, and $aF = c^2$.

239. If to a tangent of an ellipsis, a perpendicular be drawn from either focus, the distance of that focus from the point of contact will be to the distance of the other focus from the same point, as the square of the perpendicular drawn to the tangent, to the square of the semi-conjugate axis.

a. Let P (fig. 23.) be a point in the ellipsis ADBE, AB the transverse, DE the conjugate axis, C the centre, S and F the foci; GPH a tangent to the ellipsis in P, SG the perpendicular on it from S. Draw SP, PF, and make FH perpendicular to GP. Because the angles SPG, FPH, are equal, from the nature of the ellipsis, the triangles SPG FPH are equiangular, and therefore

$$SP : PF :: SG : FH, \text{ and}$$
$$SP : PF : SG^2 : FH \times SG.$$

But

But $FH \times SG = CD^2$, therefore
$SP : PF :: SG^2 : CD^2$.

b. Hence $\frac{CD^2}{SG^2} = \frac{PF}{SP}$.

240. If a body urged by a centripetal force, directed to a fixed point, describe an ellipsis of which that point is a focus, the centripetal force must be inversely as the square of the distance.

Suppose a body falling in a straight line to the focus S, (fig. 24.) to have at A the same velocity that the body revolving in the ellipsis has at A. Let P be any other position of the revolving body, and with the radius SP let an arch of a circle be described, meeting AB in L; the velocity of the falling body at L, and of the revolving body at P, will be equal. If, then, the velocity of the revolving body at the mean distance, or at the point D, be called c, and its velocity at P be v, $v : c :: CD : SG$, (§ 236, *b.*), or $v = \frac{c \cdot CD}{SG}$, and $v^2 = \frac{c^2 \cdot CD^2}{SG^2}$ $= c^2 \times \frac{PF}{SP}$, (§ 239, *a.*) Therefore also $v = c\sqrt{\frac{PF}{PS}}$.

Now

Now v is the velocity of the falling body at L, as well as of the revolving body at P; therefore if LN be taken equal to FP, the velocity of the falling body at L is $= c\sqrt{\frac{NL}{SL}}$. But because SL is equal to SP, and LN to PF, SN=SP+PF=AB, so that the point N is given. The centripetal force at L is therefore as $\frac{1}{SL^2}$, § 238., that is, as $\frac{1}{SP^2}$, or inversely as the square of the distance. Q. E. D.

241. The forces, therefore, which make the planets describe ellipses having the Sun in their common focus, are inversely as the squares of the distances from the centre of the Sun.

a. As the same reasoning that is here applied to the ellipse, might be applied to the hyperbola or the parabola, therefore the force tending to one of the foci, which is requisite to make a body describe a conic section, is inversely as the square of the distance.

b. The converse of this is true, viz. that if the centripetal force be inversely as the square of the distance, the body will describe a conic section.

c. The

c. The line NS, (fig. 24.) through which a body falling toward S will have, at any distance from S, the same velocity that the revolving body has at that same distance, is equal to the transverse axis of the ellipsis; and the velocity of the revolving body at D, its mean distance, is equal to that which is acquired by falling from N to K, the middle point between N and S. If d be a given distance from S, and f the centripetal force at that distance, because $\frac{1}{d^2} : \frac{1}{a^2} :: f : F$, $f = \frac{a^2 F}{d^2}$ or $d^2 f = a^2 F$.

But $a F = c^2$, therefore $d^2 f = a c^2$ and $c = \frac{d f^{\frac{1}{2}}}{a^{\frac{1}{2}}}$.

242. If bodies describe different ellipses about a centre to which they are urged by centripetal forces in the inverse ratio of the squares of the distances, the squares of the times of revolution will be as the cubes of the mean distances.

The velocity c, at the mean distance a, is $\frac{d f^{\frac{1}{2}}}{a^{\frac{1}{2}}}$, and therefore the sector described in a second by the radius vector, when the body is at its mean distance, is $\frac{b c}{2} = \frac{b d f^{\frac{1}{2}}}{2 a^{\frac{1}{2}}}$, and as oft as this is contained

tained in the whole elliptic area, of so many seconds does the time of the revolution consist. Now the area of the ellipsis of which a is the semi-transverse, and b the semi-conjugate, is $\pi\, a\, b$; therefore the

$$\text{time of revolution} = \pi\, a\, b \times \frac{2a^{\frac{3}{2}}}{b\, d f^{\frac{1}{2}}} = \frac{2\,\pi\, a^{\frac{3}{2}}}{d f^{\frac{1}{2}}}.$$

If, therefore, t and t' be the times of the revolution of two planets,

$$t : t' :: a^{\frac{3}{2}} : a'^{\frac{3}{2}}, \text{ or}$$

$$t^2 : t^{2\prime} :: a^3 : a'^3.$$

The third and last of the laws of KEPLER, is thus derived from the action of a centripetal force, obeying the law already shewn to be conformable to the other two, that is, being inversely as the squares of the distances.

243. Suppose a body to gravitate to a centre, as in the preceding proposition, and to be projected from a given point, with a given velocity, in a direction perpendicular to the line drawn to the centre; required to find the Conic Section which it will describe.

a. Let the semi-transverse or the mean distance $= a$; the semi-conjugate $= b$; the velocity at the mean distance $= c$; and at the apsis A, (fig. 24.) from which the body is projected, $= v$; let the line SA $= r$.

From

From (§ 236, *b*.) $b\,c = r\,v$; now, $c = \dfrac{d f^{\frac{1}{2}}}{a^{\frac{1}{2}}}$; therefore

$b\,d f^{\frac{1}{2}} = r\,a^{\frac{1}{2}}\,v$, or $b^2 d^2 f = a\,v^2 r^2$; and because

$$b^2 = \text{AS} \times \text{SB} = r\,(2\,a - r),$$

$$r\,(2\,a - r)\,d^2 f = a\,v^2 r^2, \text{ and}$$

$$a = \frac{r\,d^2 f}{2\,d^2 f - v^2 r}.$$

Thus the semi-transverse axis is found, and from it, with the focus S, and the apsis A, the conic section may be described.

b. The conic section will be a circle, when $a = r$, or, when $d^2 f = v^2 r$, or $v^2 = \dfrac{d^2 f}{r}$. If v is such that $2\,d^2 f = v^2 r$, or $v^2 = \dfrac{2\,d^2 f}{r}$, the denominator is 0, and the value of a becomes infinite, so that the trajectory is a parabola, of which the focus is *s*, and the parameter $4\,r$.

c. When $2\,d^2 f > v^2 r$, the value of a is affirmative, and the conic section is an ellipsis; and this ellipsis has its higher apsis at A, if $v^2 < \dfrac{d^2 f}{r}$; but when v^2 is between the limits of $\dfrac{d^2 f}{r}$ and $\dfrac{2\,d^2 f}{r}$, the lower apsis is at A.

d. When

d. When v goes beyond this latter limit, or when $v^2 r > 2 d^2 f$, the value of a is negative, and the trajectory becomes a hyperbola.

244. From the fact that action is always accompanied by re-action, we conclude that gravitation among terrestrial bodies is the mutual tendency of the particles of matter to one another. It is therefore reasonable from analogy to suppose, that this is true in all cases, and that the force of gravitation toward different bodies, the distances being the same, is proportional to the quantities of matter, or the masses of the bodies. If the mass of a body be called m, the gravitation to it at any distance x will be $\frac{m}{x^2}$.

Hence, in the formulas above, m may be inserted instead of $d^2 f$.

245. The quantities of matter in any two primary planets, are directly as the cubes of the mean distances at which their satellites revolve, and inversely as the squares of the periodic times of those satellites.

a. This is proved by substituting $m^{\frac{1}{2}}$ for $df^{\frac{1}{2}}$, in the formula $t = \frac{2 \pi a^{\frac{3}{2}}}{df^{\frac{1}{2}}}$; we have thence

$$t : t' :: \frac{a^{\frac{3}{2}}}{m^{\frac{1}{2}}} : \frac{a'^{\frac{3}{2}}}{m'^{\frac{1}{2}}},$$

from which the proposition follows readily.

b. By means of this theorem, the masses of the Sun, and of the planets which have satellites, may be compared with one another.

c. In the *Mecanique Celeste*, they are calculated from the most exact data, as below:

Quantity of matter in the Sun being $= 1$

That in the Earth, $= \frac{1}{329630}$

—— in Jupiter, $= \frac{1}{1067.09}$

—— in Saturn, $= \frac{1}{3594}$

—— in Uranus, $= \frac{1}{19504}$

d. As the ratios of the Diameters of the planets are known from observation, the ratios of their Bulks, being the same with those of the cubes of their diameters, are also known; and hence the Densities, which are proportional to the quantities of matter, divided by the bulks, are found.

Density

Density of the Sun,	=	1
——— of the Earth,	=	3.9393
——— of Jupiter,	=	0.8601
——— of Saturn,	=	0.4951
——— of Uranus,	=	1.1376

246. The immoveable point to which the planets gravitate, is not the centre of the Sun, but the centre of gravity of the solar system.

From the equality of action and re-action, the gravitation of the planets to the Sun must be accompanied by the gravitation of the Sun to the planets, so that the quantity of the motion of the former, estimated in any direction, must be equal to that of all the latter estimated in the opposite. The Sun, therefore, moves in an orbit, about the only point of which the condition cannot be disturbed by the mutual action of the surrounding bodies, viz. the centre of gravity of the whole.

If there were only one planet, the Sun and that planet would describe similar conic sections, of which their common centre of gravity would be one of the foci; their distances from that point being always inversely as their masses. If there be a number of planets, the path of the Sun will become a more complicated curve, but will be such as to furnish a centrifugal force in respect of each planet, just able to counteract the gravitation toward it.

247. The centre of the Sun is never distant by so much as *his own diameter from the centre of gravity of the system.*

The diameter of the Sun is equal nearly to .009 of the radius of the Earth's orbit. Now, if we suppose the Sun, and all the great planets of the system, Jupiter, Saturn and Uranus, to be in a straight line, and the planets all on one side of the Sun, the centre of the Sun will be nearly the farthest possible from the centre of gravity of the whole; yet we shall find on computation, that the distance is not greater than .0085 of the radius of the Earth's orbit. NEWTONI *Princip.* lib. III. prop. 12.

248. In the preceding propositions the existence of the principle of gravitation, has been established by induction from the laws of KEPLER; and from that principle, by reasoning downward, conclusions have been obtained concerning the quantity of matter in the planets, to which observation, without the assistance of theory, never could have reached.

It yet remains to be shewn, that the same force which occasions the descent of heavy bodies on the Earth's surface, at the rate of 16.09 feet *per second*, when diminished in the inverse ratio of the square of the distance, is just sufficient to retain the Moon in her orbit.

The

The siderial revolution of the Moon is 27d.32166, § 124. and her mean distance 59.879 semidiameters of the Earth. From this, on the supposition that the squares of the periodic times are as the cubes of the distances, the period of a body projected, so as to describe a circle round the Earth, will be found 1h.41516. From this the arch which the body describes in one second, is found to be 4′.13″.48‴; and the deflection from the tangent in one second, if reduced to feet, comes out 16 and a small fraction, the same with the descent of a heavy body in one second, at the surface of the Earth.

NEWTONI *Princip.* lib. III. prop. 4.

On the subject of this section, see NEWTONI *Princip. Math.* lib. 1mus, sectio 2da, 3tia, 7ma, 8va, 11ma.

JOH. BERNOUILLI, *Opera*, tom. 1mus, N°. 86.

EULERI *Mechanica*, tom. I. cap. 5tum. Particularly prop. 80, 81.

FRISII *Opera*, tom. 3tius, *Cosmographia*, lib. 1mus.

Dr MATHEW STEWART, *Tracts Phys. & Math.* Tract 1. The subject is there treated in a manner strictly geometrical.

LA LANDE, liv. 2de, tom. III.

VINCE, *Astron.* chap. 31. vol. II.

Theorie des Mouvemens des Planetes, par M. LA PLACE, 1re partie, Paris, 1784.

SECT.

SECT. II.

OF THE FORCES WHICH DISTURB THE ELLIPTICAL MOTION OF THE PLANETS.

WHEN there are only two bodies that gravitate to one another, with forces inversely as the squares of their distances, it appears from the last section that they move in conic sections, and describe, about their common centre of gravity, equal areas in equal times, that centre either remaining at rest, or moving uniformly in a straight line. But if there are three bodies, the action of any one on the other two, changes the nature of their orbits, so that the determination of their motions becomes a problem of great difficulty, distinguished by the name of THE PROBLEM OF THE THREE BODIES.

The solution of this problem, in its utmost generality, is not within the power of the mathematical sciences, as they now exist. Under certain limitations, however, and such as are quite consistent with the condition of the heavenly bodies, it admits of being resolved. These limitations are, that the force which one of the bodies exerts on the other two, is, either from the smallness of that body, or its great distance, very inconsiderable, in respect

respect of the forces which these two exert on one another.

The force of this third body is called *a disturbing force*, and its effects in changing the places of the other two bodies are called *the disturbances of the system*.

Though the small disturbing forces may be more than one, or though there be a great number of remote disturbing bodies, the computation of their combined effect arises readily from knowing the effect of one; and therefore the problem of Three bodies, under the conditions just stated, may be extended to any number.

Two very different methods have been applied to the solution of this problem. The most perfect is that which embraces all the effects of the disturbances at once, and by reducing the momentary changes into fluxionary or differential equations, proceeds, by the *integration* of these, to determine the whole change produced in any finite time, whether on the angular or rectilineal distance of the bodies. This method gives all the inequalities at once, and as they mutually affect one another.

The other method of solution is easier, and more elementary, but much less accurate. It supposes the orbit disturbed to be nearly known, and proceeds to calculate each inequality by itself, independently of the rest. It cannot, therefore, be exact,

exact, and gives only a first approximation to the quantities sought; but being far simpler than the other, it is much better suited to the elements of science. It is also the original method, and that which was first applied by Sir ISAAC NEWTON to explain the irregularities of the Moon's motion. The same has been followed and improved by CALENDRINI, in his *Commentary on the third Book of the Principia*; by FRISI in his *Cosmographia*; and by VINCE, in the second volume of his *Astronomy*.

The other method was not invented till several years later, when it occurred nearly about the same time to the three first geometers of the age, CLAIRAUT, EULER, and D'ALEMBERT. It was followed also by MAYER, and several others, but particularly by LA PLACE, who, in the *Mecanique Celeste*, has given a complete investigation of the inequalities both of the primary and secondary planets.

I shall explain the resolution of the forces that is in some measure common to both methods; and shall shew how their effects are to be estimated in some simple instances, going from thence to the enumeration of the results. I begin with the Moon's irregularities, as the easiest case of the problem.

249. The Moon, in her motion round the Earth, is disturbed by the action of the Sun; her gravity to the Earth is increased near the quadratures,

quadratures, and diminished both at the opposition and the conjunction, and the areas described by the radius vector, except near the quadratures, are never exactly proportional to the times.

a. Let ADBC (fig. 25.) be the orbit, nearly circular, in which the Moon M revolves round the Earth at E. Let the Sun be at S, and let the line SE denote the force of the Sun on the Earth; then $\frac{SE^3}{SM^2}$ is the attraction of the Sun on the Moon at M; let MG be equal to this line, and joining ME, let the parallelogram KF be described, of which GM is the diagonal, and the sides MF, MK are in the directions of ME and ES; and let HS be drawn parallel to ME.

The force MG may be resolved into the two MF, MK, of which MF, directed toward the centre E, increases the gravity of the Moon to the Earth, and does not hinder the areas described by the radius vector from being proportional to the times.

The other force MK draws the Moon in the direction of the line joining the centre of the Sun and Earth; but it is only the excess of this force above that by which the Sun draws the Earth, that disturbs the relative position of the Moon and Earth. If, therefore, ES, which denotes the force of the Sun on the Earth be taken from MK, the remainder KH will be the force by which the Sun draws

draws the Moon from the Earth, in the direction of the line passing through the centres of the Sun and Earth.

This force at the conjunction, exceeding that part of the disturbing force which draws the Moon to the Earth, tends to diminish the Moon's gravity to the Earth. At the opposition B it does the same, by becoming negative; for the Earth is then drawn more than the Moon, and the difference is nearly the same as at the conjunction.

b. If MN be taken equal to HK, and NO made perpendicular to the radius vector, the force MN is resolved into two, one directed from M to O, lessening the gravity of the Moon to the Earth, and the other directed from O to N, parallel to the tangent to the Moon's orbit at M, and therefore accelerating the Moon from C to A, retarding her from A to D, and so alternately in the other two quadrants.

At the quadratures C and D, the force HK vanishes, and the only remaining force is directed to the centre of the Earth, so that the areas are there proportional to the times.

c. The analytical values of these forces are next to be found. Draw CED the line of the quadratures, put SE $= a$, EM the radius vector of the Moon's orbit $= r$, the angle CEM $= x$, and the mass of the Sun $= m$. The force that retains the Earth in its orbit

orbit $= \frac{m}{a^2}$; also the Sun's force on M in the direction SM $= \frac{m}{(a - r \sin x)^2}$ nearly, and hence the force HK, or MN, is found $= \frac{m}{a^3} \cdot 3 r \sin x$.

Therefore, MO $= \frac{m}{a^3} \cdot 3 r \sin^2 x$, and

NO $= \frac{m}{a^3} \cdot 3 r \sin x \times \cos x = \frac{m}{a^3} \cdot \frac{3 r}{2} \sin 2x$.

Also MF $= \frac{m r}{a^3}$.

In all this the terms involving powers of r greater than the first are rejected; and hence also, MF — MO, or the whole force with which the gravity of the Moon to the Earth is increased or diminished, is $\frac{m r}{a^3}(1 - 3 \sin^2 x)$. This last force is 0, when $3 \sin^2 x = 1$, or $\sin x = \frac{1}{\sqrt{3}}$, that is, when $x = 35°.15'.5''$; at the quadratures the same force is $+ \frac{m r}{a^3}$, and at the opposition and conjunction $- \frac{2 m r}{a^3}$.

The

The mean quantity of the force $\frac{m\,r}{a^3}(1 - 3 \sin x^2)$, for an entire revolution, is found by multiplying it by $\dot{x}$, when it becomes $\frac{m\,r}{a^3}(\dot{x} - 3\,\dot{x} \sin^2 x)$, the fluent of which, or the sum of the disturbing forces =

$$\frac{m\,r}{a^3}(x - \frac{3}{2}\,x + \frac{3}{2} \sin x \times \cos x);$$

and this, when x is an entire circumference, becomes $-\frac{m\,r}{a^3} \times \frac{\pi}{2}$. This must be divided by π, and it gives $-\frac{m\,r}{2\,a^3}$ for the mean disturbing force acting on the Moon, in the direction of the *radius vector*.

250. Hence, at the quadratures, the gravity of the Moon to the Earth is increased by a quantity equal to the mass of the Sun multiplied into the radius of the Moon's orbit, and divided by the cube of the Sun's distance from the Earth; at the opposition and conjunction it is diminished by twice this quantity; and the effect

fect, upon the whole, is to diminish the Moon's gravity, by one-half of the said quantity, or by $\frac{m r}{2 a^3}$.

The disturbing force in the direction of the radius vector may be more conveniently expressed; for if F be the force by which the Earth attracts the Moon, P the periodical time of the Earth's revolution round the Sun, and P′ that of the Moon round the Earth, it has been already shewn, that

$$\frac{m}{a^2} : F :: \frac{a}{P^2} : \frac{r}{P'^2},$$

$$\text{and } \frac{m r}{a^3} : F :: \frac{1}{P^2} : \frac{1}{P'^2} : P'^2 : P^2.$$

$\frac{m r}{a^3}$ is therefore nearly $\frac{1}{179}$ of the Moon's gravity, and the mean disturbing force $= \frac{1}{358}$.

251. From the diminution of her gravity by a 358th part, the Moon describes her orbit at a greater distance from the Earth, with a less angular velocity, and in a longer time, than if she were urged to the Earth by her gravity alone.

If the solar action were to cease, the Moon would come nearer to the Earth, and would describe her orbit

orbit in less time. As the area described by the radius vector in the primitive and the disturbed orbit is the same, it can be shewn, that in consequence of the mean disturbing force, the radius vector is increased by a 358th part, and the angular velocity diminished by a 179th part.

Exposition du Systeme du Monde, chap. v. p. 213. 2de edit.

252. The annual equation is an irregularity in the Moon's motion, arising from the variation of the Sun's distance from the Earth, and bearing a given ratio to the equation of the Sun's centre.

Since the Sun's disturbing force is inversely as the cube of his distance, when he approaches the Earth at the perihelion, the Moon's motion is slower, and, for the same reason, at the aphelion, it is quicker than the mean. This produces what is called the *annual equation*, equal nearly to $(11'.12'') \times \sin$ Sun's mean anomaly. It is nearly equal to one-twelfth part of the equations of the Sun's centre, but has the opposite sign, being subtracted when that equation is added, and *vice versâ*.

253. An equality in the same disturbing force, depending on the position of the transverse axis of the Moon's orbit, in respect of the line drawn from the Earth to the Sun, produces the equation

tion already mentioned by the name of the *Evection*, § 129. *b*.

When the transverse axis of the Moon's orbit is directed to the Sun, or lies in the line of the syzygies, because the force, taken in syzygies from the gravitation of the Moon, is, *cæteris paribus*, proportional to her distance from the Earth; it is greatest when the Moon is in the apogee, and least when the Moon is in the perigee; the greatest possible diminution is therefore made from the quantity that is already the least, and the least diminution from that which is already the greatest; their difference therefore is augmented, and the motion at the apogee made slower than usual, and at the perigee faster. When the line of the apsides is in the quadratures, the contrary happens; the gravitation at the apogee is most augmented, and at the perigee least augmented: the difference therefore is diminished.

In the former case, the equation of the centre appears to be increased, and in the second it appears to be diminished. The quantity is — $(1^\circ.20'.36'')$ multiplied into the sine of twice the Moon's angular distance from the Sun, *minus* the mean anomaly of the Moon. Mayer adds another term to the evection, equal to $+ 26''$ multiplied into the sine of double the preceding argument. *Theoria Lunæ*, Lond. 1767, § 50.

254. The

254. The *Variation* is an equation derived from the force at right angles to the radius vector, which, from the quadratures to the syzygies, accelerates the motion of the Moon, and from the syzygies to the quadratures retards it. This force was found to be proportional to the sine of twice the Moon's angular distance from the Sun, and the equation itself is nearly proportional to the same quantity.

If the angular distance of the Moon from the Sun $= \Delta$, the variation, according to MAYER, is =

$$+ (35'.47'') \sin 2\Delta$$
$$+ (2'') \, 3 \sin \Delta$$
$$+ (14'') \sin 4\Delta.$$

Beside the above inequalities, first found out by observation, and now explained by the Theory of Gravity, there are several others of smaller amount, which theory alone has discovered; all that observation could do, being to ascertain that some unknown inequalities existed, which introduced an uncertainty into all the calculations of the Moon's place. The causes of this uncertainty have been discovered by the theory, and their effects calculated, so that the Tables of the Moon have been brought to agree very nearly with observation.

255. One of the equations which theory has discovered, is remarkable for the great length of

of its period, being no less than 85 years. It depends on the position of three lines; the axis of the Moon's orbit, the axis of the Earth's orbit, and the line of the Moon's nodes. If N be the longitude of the node, P that of the Moon's perigee, and P′ that of the Sun's; this equation = 14″ sin (2 N + P — 3 P′.)

Mecanique Celeste. Also Preface to the New Tables, VINCE's *Astron.* tom. III.

256. The inequality called the *acceleration of the Moon,* by which her velocity appears subject to continual increase, and her period to continual diminution, has been found by LA PLACE to be a *secular* equation, (depending on a change in the eccentricity of the Earth's orbit), which does not run through the circle of its changes but in the course of several thousand years. For many centuries to come it may be nearly expressed by this formula, where n denotes the number of centuries from 1700,

$$+ 10''.181621268 \times n^2 + 0''.0185384408\, n^3.$$

The other lunar inequalities respect the orbit itself, where, in consequence of the Sun's disturbing force, these three *elements* are subject to change, viz. The position of the line in which the plane of the orbit cuts the plane of the ecliptic; also the angle in which it cuts that plane; and, lastly, the direction

tion of the transverse axis of the orbit, or of the line of the apsides.

257. The motion of the line of the nodes is produced by that part of the Sun's disturbing force which is in the direction of the straight line joining the centres of the Sun and Earth, and proportional to the distance of the Moon from a plane passing through the centre of the Earth at right angles to the line joining the centres of the Sun and Earth.

Suppose the Moon to be on the same side of the last-mentioned plane that the Sun is. Then, if in the direction of her motion for the instant just past, there be taken a line equal to the space passed over in that instant, and if in the line drawn through the Moon, perpendicular to the above plane, on the side opposite to the plane, there be taken a part equal to the space which the force urging the Moon from the plane would have made her describe in the same time, then the true path of the Moon will be the diagonal of the parallelogram under these two lines; and the momentary change in the place of the node will be the distance between the point where an arch having the direction of this diagonal, and another having the direction of its side, meet the plane of the ecliptic. The same will happen in other situations of the Moon, and the line of the nodes will thus have a motion in in a contrary direction to the motion of the Moon, that is, a retrograde motion, or *in antecedentia.*

258. From this, NEWTON has shewn, that the horary motion of the node at any time is to 33″.177, as the product of the sines of these three angles, the distance of the Moon from the quadrature, the distance of the Moon from the node, and the distance of the node from the Sun, to the cube of the radius.

Princip. lib. III. prop. 31.

When any one of the above three sines becomes negative, the motion of the node, from being retrograde, becomes progressive.

259. The mean horary motion of the node, which arises from taking the average of all the horary motions, is to 16″.2771 as the square of the sine of the distance of the node from the Sun to the square of the radius.

Ibid. cor.

Hence the mean motion of the nodes for a siderial year is 19°.18′.1″.4, which is not far from the result of astronomical observation.

260. From the same cause arises a variation in the inclination of the Moon's orbit to the ecliptic, confined within very narrow limits. A construction for determining the inclination for any given time, is given *ibid.* prop. 35., and is found to agree with observation.

What respects the motion of the nodes is thus completely explained; and it is here that the *indirect* method of determining the Moon's inequalities has been most successful. It has not been equally so in ascertaining the motion of the apsides.

To conceive, in general, the cause which renders the apsides of the Moon's orbit more than 180 degrees distant from one another, we must begin with supposing the Moon at the *lower apsis;* then, if that planet were acted on only by the force of gravity, the radius vector, after it had described 180°, would arrive at the *upper apsis*, or would be intersected by the orbit at right angles.

But as the mean disturbing force, in the direction of the radius vector, may be considered as a quantity constantly taken from the Moon's gravity, the portion of her path described in any instant will fall between the tangent and the arch of the elliptic orbit which would have been described if the Moon had been acted on by gravity alone. The actual path of the Moon, therefore, will be less bent than the elliptic orbit would have been in the same

same time, and will not come to intersect the radius vector at right angles, till this last has moved over a greater arch than 180°. Hence the solar force, by lessening the Moon's gravitation to the Earth, produces an advance in the place of the apsides, in the direction of the Moon's motion. If the same force had been added, it would have produced a motion *in antecedentia*.

The precise quantity of the motion of the apsides is not, however, easily determined. NEWTON left this part of the theory almost untouched; and the only theorem which he gave, having any reference to it, made the motion only half of what it is found in reality to be. MACHIN was, I believe, the first, who, after NEWTON, attempted this investigation; he has only mentioned the result, which is very exact, and the principles on which his reasoning was founded, viz. that from the disturbing force in the direction of the radius vector, he determined the Moon's nearest approach to the Earth, and farthest recess from it, supposing that but for such disturbance, the Moon's orbit would have been a circle: from this also he found the time from the lowest to the highest apsis.

This method was afterwards adopted by Dom. WALMSLEY, and by Dr MATHEW STEWART, who both derived from it the true motion of the apsides, by investigations extremely ingenious, and that

that of the latter, particularly distinguished for its geometrical elegance. Dr STEWART has demonstrated this theorem:

261. If r be the radius of the Moon's orbit, supposing it to be a circle, and acted on only by F, her gravity to the Earth. Then if a mean disturbing force $= f$ be supposed to diminish the Moon's gravity, the greatest distance she will go to from the Earth is $r \times \frac{F - 3f}{F - 5f}$; and the cube of this distance will be to the cube of r in the duplicate ratio of the angle described by the Moon from one apsis to the next to two right angles.

Tracts Math. & Phys. Tract IV. prop. 27.; also *Sun's Distance*, prop. 9.

Hence the angle described from one apsis to the next $= \frac{(F - 3f)^{\frac{3}{2}}}{(F - 5f)^{\frac{3}{2}}} \times 180°$.

FRISI has adopted the principle of MACHIN, including also the action of the forces perpendicular to the radius vector. *Opera*, tom. III. p. 380.

VINCE, in the 2d volume of his *Astronomy*, has computed the motion of the apsides according to this last method, and makes the quantity of it in a siderial

derial revolution of the Moon, $3^{\circ}.2'.32''.39$; and in a year $40^{\circ}.40'.20$.

MAYER'S *Tables* make the latter $40^{\circ}.41'.33''$. Dr STEWART'S *Theorem* gives $3^{\circ}.1'.20''$, supposing $\frac{f}{F} = \frac{1}{358}$, the value assigned to it, § 250.

The result of these investigations, therefore, agrees nearly with observation; but it cannot be denied that the principle on which they are founded is liable to some objections, so that if it were not for the information derived from the direct solution of the problem of *the three bodies*, it might still be doubted, whether the principle of gravity accounted exactly for the motion of the Moon's apsides. CLAIRAUT, who first compared the result of that solution with observation, met with the same difficulty that NEWTON had done, and found that his formula gave only half the true motion. He therefore imagined that gravity is not inversely as the squares of the distances, but follows a more complicated law, such as can only be expressed by a formula of two terms. In seeking for the co-efficient of the second term, he was obliged to carry his approximation farther than he had done before; in consequence of which, the co-efficient he sought for came out equal to 0, and the motion of the apsides was found to be completely explained by the supposition,

position, that the force of gravity is inversely as the square of the distance.

262. Since, by the Moon's inequalities, the ratio of the Sun's disturbing force to the force with which the Moon gravitates to the Earth becomes known; if the ratio of the former, to the whole force retaining the Earth in its orbit, could be found, the ratio of this last to the force of the Earth on the Moon would also be found, and from thence the ratio of the distance of the Moon to the distance of the Sun from the Earth.

The latter ratio would be given, because if F and f are the forces that retain the Earth and the Moon in their orbits; if a and r are the radii of those orbits, P and p the periodic times,

$$F : F' :: \frac{a}{P^2} : \frac{r}{p^2};$$

and so if the ratio of F to F′ be given, P and p being also given, the ratio of a to r is found.

In this way the Sun's distance might be found from his own force to disturb the Moon, the idea of which appears first to have occurred to Dr MATHEW STEWART. The principal difficulty is to find the ratio of the disturbing force of the Sun to the force which the Sun exerts on the Earth. If the expression of that force was carried

no

no farther than to involve the simple power of $\frac{r}{a}$, the investigation would lead to nothing, as a would disappear from the equation; but if it be carried to the terms that involve $\frac{r^2}{a}$, an equation may be found from which a can be determined. Dr STEWART, therefore, having first compared the mean disturbing force with the gravity of the Moon to the Earth, from the motion of the apsides, proceeded to determine the ratio of the former force to the force that retains the Earth in its orbit, by carrying the approximation to the terms that involved the square of the Sun's distance, and from that computed the Sun's parallax at 6″.9. *Sun's Distance*, prop. 10.

This value of the parallax is no doubt too small; it is only an inconsiderable part of the disturbing force of the Sun, into which his distance enters as an element, and therefore any deduction founded on it must be liable to error.

MAYER has also sought to determine the Sun's parallax from one of the lunar equations, as deduced from the solution of the problem of *the three bodies*. The co-efficient of this equation involves the Sun's parallax, and gives it equal to 7″.8, and MAYER thinks that the error cannot exceed a 24th part. *Theoria Lunæ*, § 51.

263. The

263. The Tables of the Moon's motions in the state to which they are now brought, contain 28 equations for the longitude of the Moon; 12 for the latitude, and 13 for the horizontal parallax, or the distance of the Moon from the Earth; and their greatest probable error does not exceed 12 seconds.

It is very much to the introduction of the equation of the period of 85 years, which nobody before LA PLACE had discovered, that the superior accuracy of the New Tables is to be ascribed.

The observations at Greenwich have afforded the data from which the theory of gravity, and the *integral calculus* have extracted these memorable results.

SECT.

SECT. III.

DISTURBANCES IN THE MOTIONS OF THE PRIMARY PLANETS PRODUCED BY THEIR ACTION ON ONE ANOTHER.

It is necessary, in this inquiry, to know the quantities of matter in the different planets; and these have been already calculated for the planets which have satellites. The masses of Venus and Mars have been computed by M. La Place from some disturbances which they appear to produce on the Earth's motion. The mass of Mercury has been estimated, from supposing his density, and that of the Earth, to be inversely as their mean distances from the Sun. This law holds with respect to the Earth, Jupiter and Saturn, and analogy authorises the extension of it to Mercury. From knowing the density and the bulk, the quantity of matter is inferred. The mass of the Sun being 1, that of Mercury is $\frac{1}{2025810}$, of Venus $\frac{1}{383137}$, of Mars $\frac{1}{1846082}$, those of the other being as already stated, § 245.

The

The gravitation of one planet to another, is expressed by the quantity of matter in each, divided by the square of the distance; and therefore the two bodies tend to come together with a force that is as the sum of their masses divided by the square of the distance; so that when the motion of both is referred to one only, the force must be expressed by the sum of the masses divided by the square of the distance.

The disturbances produced by the action of the primary planets on one another, are of more difficult investigation than those produced by the Sun on the motions of the Moon, because the disturbing body is not at an immense distance, as in the latter case. The only sure way of subjecting them to calculation, is by a direct solution of the Problem of the Three Bodies; the part of which that may be accounted quite elementary is now to be considered.

264. The forces which act upon a body, to however many centres they tend, and whatever law they may obey, may be resolved into the directions of three lines or *axes*, given in position, at right angles to one another.

This is evident from Dynamics, vol. I. § 70.

The advantage of this resolution of forces for determining the motion of a body attracted to several centres,

centres, appears to have been first suggested by MACLAURIN, *Fluxions*, § 470. It has been very useful in the solution of the problem of *The three bodies*.

265. Let S be the Sun, (fig. 26.), and P and P′ two planets referred to the plane of the ecliptic, each by three rectangular co-ordinates, PQ, QR, RS, and P′Q′, Q′R′, R′S, parallel to the three axes SO, SN, NM : let SR = x, RQ = y, PQ = z. Let all the forces that draw P in the direction of SM, or in a line parallel to x, whether arising from the action of the Sun S, or the Planet P′, be denoted by F; all those in the direction of SN or y by F′; and those in the direction of SO or z, by F″; and let $\dot{t}$ be the momentary increment or fluxion of the time; then

$$\ddot{x} = -\mathrm{F}\dot{t}^2$$

$$\ddot{y} = -\mathrm{F}'\dot{t}^2, \text{ and}$$

$$\ddot{z} = -\mathrm{F}''\dot{t}^2.$$

For from the principles of Dynamics, (1. § 100.)

$$-\mathrm{F}\dot{x} = v\dot{v}; \text{ and (1. § 56.) } v = \frac{\dot{x}}{\dot{t}}, \text{ so that}$$

$\dot{v} =$

$\dot{v} = \frac{\ddot{x}}{\dot{t}}$, because $\dot{t}$ is supposed constant; therefore $v\dot{v} = \frac{\dot{x}}{\dot{t}} \times \frac{\ddot{x}}{\dot{t}} = \frac{\dot{x}\ddot{x}}{\dot{t}^2}$, and hence

$$-F\dot{x} = \frac{\dot{x}\ddot{x}}{\dot{t}^2}, \text{or } F = -\frac{\ddot{x}}{\dot{t}^2},$$

that is $\ddot{x} = -F\dot{t}^2$.

In the same manner are found the values of $\ddot{y}$ and of $\ddot{z}$.

On these three equations is founded the solution of the problem of The three bodies.

266. The same things being supposed, let the co-ordinates by which the planet P′ is referred to the ecliptic be x', y', z'; let SP = $\sqrt{x^2 + y^2 + z^2} = r$, SP′ = $\sqrt{x'^2 + y'^2 + z'^2} = r'$, and the distance of the planets, or PP′ = $\sqrt{(x - x')^2 + (y - y')^2 + (z - z')^2} = q$. Also let m be the mass of the planet P, and m' that of P′, the mass of the Sun being 1. The whole force

force parallel to SM or x, that affects the relative motion of P to S, is

$$\frac{(1+m)x}{r^3} + \frac{m'x'}{r'^3} + \frac{m'\dot{q}}{q^2\dot{x}}.$$

For the action of the Sun on P being resolved as above, the part of it that is in a direction parallel to x or to SM is $\frac{x}{r^3}$; and the action of P′ on P being resolved in like manner, the part of it parallel to x is $\frac{m'(x-x')}{q^3}$. Now, supposing x to vary by the momentary increment or fluxion $\dot{x}$, x' remaining constant, the increment of q, or $\dot{q}$ will be such that $\frac{\dot{q}}{\dot{x}} = \frac{x-x'}{q^2}$, and therefore $\frac{m'(x-x')}{q^3} = \frac{m'\dot{q}}{q^2\dot{x}}$, which, therefore, is the force by which P′ acts directly on P; and if to this be added the Sun's force in the same direction, viz. $\frac{x}{r^3}$, the amount of the *direct* action of S and P′ on P, in a line parallel to SM, is $\frac{x}{r^3} + \frac{m'\dot{q}}{q^2\dot{x}}$.

Now,

Now the Sun, by the united action of the planets P and P′, is drawn in the direction opposite to this last by a force $= \frac{m x}{r^3} + \frac{m' x'}{r'^3}$, and as the Sun is here considered as immoveable, we must conceive this force to be transferred to P in the opposite direction.

Thus the whole action on P, or the force F =

$$\frac{x}{r^3} + \frac{m x}{r^3} + \frac{m' x'}{r'^3} + \frac{m' \dot{q}}{q^2 \dot{x}}$$

$$= \frac{(1+m) x}{r^3} + \frac{m' x'}{r'^3} + \frac{m' \dot{q}}{q^2 \dot{x}}.$$

By writing in this formula y, y' and $\dot{y}$, instead of x, x' and $\dot{x}$, we have $F' = \frac{(1+m) y}{r^3} + \frac{m' y'}{r'^3} + \frac{m' \dot{q}}{q^2 \dot{y}}$;

and in like manner

$$F'' = \frac{(1+m) z}{r^3} + \frac{m' z'}{r'^3} + \frac{m' \dot{q}}{q^2 \dot{z}}.$$

The substitution of these values of F, &c. in the three formulas of the last article, will give three fluxionary equations, on which the motion of P depends.

The same being done for P′, there will come out six fluxionary equations, from the integration of which

which the motions both of P and P′ may be determined.

This is the problem of the Three Bodies, as far as the principles of Dynamics are concerned. The rest depends on the integral calculus, and involves a multitude of investigations which do not belong to this place. See Memoir of LA PLACE, *sur les Inegalités de Jupiter et Saturn. Acad. des Sciences*, 1785. Also *Mecanique Celeste*, liv. VI.

What is given above, is merely an elementary exposition of the dynamical principles employed in these investigations; a sketch follows of some of the most important conclusions deduced from them.

267. The place of every planet in its orbit, is changed by the action of the other planets, and the orbit itself is changed in all its elements but two, the mean distance from the Sun and the mean motion of the Planet.

a. In the orbit of a planet, the line of the nodes, the inclination to the plane of the ecliptic, the line of the apsides, and the eccentricity, all vary. The lines of the nodes and of the apsides revolve continually; but the inclination of the orbit and its eccentricity are only subject to small periodical variations on each side of a mean, from which they never depart far.

b. The

b. The quantity and direction of these motions, as determined by the theory of gravity, agree entirely with the results of observation.

268. Mercury is so near the Sun, that his place in his orbit is not sensibly affected by the action of the other planets ; but his orbit is nevertheless disturbed, both in its form and position.

a. The disturbances of the four elements of the orbit are as in the following Table, where, instead of the secular change of the eccentricity, that of the greatest equation of the centre is set down. The first column gives the planet, to which the disturbances opposite to it in the succeeding columns are to be ascribed ; the second is the *annual* motion of the nodes ; the third of the apsides, both in respect of the fixed stars ; the fourth contains the *secular* variations of the inclination ; the fifth those of the greatest equation of the centre.

Action of	Node.	Aphel.	Inc. Orb.	Eq. Cent.
♀	— 5″.57	+ 4″.14	+ 3″.04	+ 9″.46
☉	— 0 .87	+ 0 .84	+ 0 .58	
♂	— 0 .14	+ 0 .04	— 0 .22	+ 0 .86
♃	— 2 .18	+ 1 .56	— 1 .26	+ 9 .87
♄	— 0 .12	+ 0 .08	+ 0 .02	+ 1 .04
	— 8 .88	+ 6 .66	+ 2 .16	+20 .43

The

The sign —, when applied to the node, or the aphelion, implies that the motion is retrograde.

b. The motions of the node and of the apogee, given in Astronomical Tables, are different from the above, as the annual procession of the equinoxes is included. Thus the annual motion of the node is — 8″.98 + 50″.25 = + 41″.27, in respect of the equinoxes, though it is only — 8″.98 in respect of the fixed stars.

269. In like manner, the place of *Venus* in her orbit is not sensibly disturbed, but the orbit itself is subject to change.

	Node.	Aphel.	Inc. Orb.	Eq. Cent.
By ☿	+ 0″.16	— 4″.30	+ 1″.91	— 9″.02
♀	— 7 .46			
☉	— 6 .69	— 5 .06		— 9 .02
♂	— 0 .29	+ 1 .18	— 0 .42	— 0 .64
♃	— 5 .13	+ 6 .38	+ 2 .00	— 6 .16
♄	— 0 .09	+ 0 .08	, 0 .35	— 0 .14
	—14″.70	+18″.40	+ 4″.47	—24″.98

270. In the orbit of the *Earth*, the apsides move forward annually at the rate of 11″.807719 in respect of the fixed stars; and the eccentricity diminishes, so that the secular variation of the greatest equation of the centre is — 17″.66.

a. Of the motion of the apsides, 5″ is due to Venus, 1″.2 to Mars, and 5″.5 to Jupiter nearly.

b. In the secular diminution of the equation to the centre, + 4″.18 is the effect of Venus, — 4″.94 of Mars, — 16″.02 of Jupiter, and the rest is produced by Mercury and Saturn.

271. As it is not the centre of the Earth, but the centre of gravity of the Moon and Earth, which describes equal areas, in equal times, about the centre of the Sun, the regularity of the Earth's motion is disturbed on that account, and the Earth is forced out of the plane of the ecliptic.

a. The irregularities thus communicated to the Earth are, by observers on its surface, transferred to the Sun; the Sun, therefore, has a motion in longitude, by which he alternately advances before the point that describes the elliptical orbit in the heavens, and falls behind it; and in like manner alternately ascends above the plane of the ecliptic, and descends below it.

b. These inequalities are small. The mass of the Moon is about $\frac{1}{68}$th of that of the Earth; the distance, therefore, of the centre of gravity of the Moon and Earth, from the centre of the latter, must be less than a semidiameter, and therefore the inequality in the Sun's place must be less than his horizontal parallax.

c. The

c. The greatest latitude which the Sun can have, is equal to the horizontal parallax, multiplied into the sine of the Moon's greatest latitude. This can hardly amount to a second; it is called the *menstrual parallax*, and was first mentioned by Mr SMEATON, *Phil. Trans.* 1768. See also *Mecanique Celeste*, tom. III. p. 106.

272. The place of *Mars* in his orbit is sensibly affected by the action of Venus, the Earth, and Jupiter.

The principal inequality produced by Jupiter is

$$-25''.6 \sin(\text{Long.}\ ♂ - \text{Long.}\ ♃)$$
$$+16''.8 \sin(\text{Long.}\ ♂ - \text{Long.}\ ♃);$$

By the Earth, $7''.2 \sin(\text{Long.}\ \odot - \text{Long.}\ ♂)$;
and by Venus, $5'.7 \sin(\text{Long.}\ ♀ - 3\ \text{Long.}\ ♂)$.

There are, besides these, some other small equations. See DE LAMBRE'S *Tables in* VINCE'S *Astronomy*, vol. III. p. 48., &c.

In the orbit of Mars, the eccentricity is diminishing. The secular variation of the greatest equation of the centre is $-37''$.

The annual and siderial motion of the aphelion is $16''.75$ forward, and of the node $23''.25$ backward. These motions are chiefly produced by the three planets just named.

273. The

273. The inequalities of the small planets *Vesta*, *Juno*, *Ceres* and *Pallas*, have not yet been computed; the disturbances which they must suffer from Mars and Jupiter are no doubt considerable, and, on account of their vicinity, though their masses are small, they may somewhat disturb the motions of one another. Their action on the other bodies in the system is probably insensible.

As two of these planets have nearly the same periodic time, they must preserve nearly the same distance, and the same aspect with regard to one another. This offers a new case in the computation of disturbing forces, and may produce equations of longer periods than are yet known in our system.

274. The action of *Jupiter* and *Saturn* on one another, produces an inequality in the motion of each, of considerable amount, and of a long period, viz. 918.76 years.

a. If n express a number of years reckoned from the beginning of 1750, S the mean longitude of Saturn, and I of Jupiter, reckoned from the same time, then the great equation that must be applied to the mean longitude of Jupiter, or to I, is

$$+ (20'49''.5 - n \times 0''.042733) \times$$
$$\sin(5\,S - 2\,I + 5^{\circ}.34'.8'' - n \times 58''.88);$$

and

and that which must be applied to S is

$$-(48'.44'' - n \times 0''.1) \times$$
$$\sin(5\,S - 2\,I + 5^{\circ}.34'.8'' - n \times 58''.88).$$

These equations are to one another nearly in the ratio of 3 to 7. As the quantity $5\,S - 2\,I - n \times 58''.88$, requires 918.76 years to increase from 0 to 360 degrees, therefore the above equations require that period to run through all their changes. See LA PLACE, *Mém. Acad. des Sciences*, 1785, 1786. Also LA LANDE, *Astron.* tom. III. § 3670.

275. Besides these two great inequalities, there are ten others, arising from the action of Saturn, to which Jupiter is subject, and which may amount when greatest to 11′56″; there are also six to which Saturn is subject from the action of Jupiter, and these may amount to 19′57″.

For the particular forms of these equations, see LA LANDE, *ibid.* and VINCE, *Astron.* vol. III. p. 94. and 102.

276. The motion of the apsides, and the change of eccentricity in the orbits of Jupiter and Saturn, are chiefly produced by the action of those planets on one another; but in the disturbance which the planes of their orbits suffer, the other planets have a sensible effect.

JUPITER.

a.

	Node.	Aphel.	Inc.	Eq. Cent.
☿	— 0″.31		— 0″.95	
♀	—17 .56	+ 0″.01	—17 .67	
☉	— 0 .01	+ 0 .01		
♂	— 0 .39		— 1 .06	— 0″.02
♃	— 6 .95			
♄	+ 5 .88	+ 6 .56	— 7 .51	+56 .28
	—19″.34	+ 6″.58	—27″.11	+56″.26

SATURN.

b.

	Node.	Aphel.	Inc.	Eq. Cent.
☿	— 0″.11		— 1″.1	
♀	— 8 .66		—26 .65	
☉				
♂	— 0 .14		— 1 .25	
♃	—12 .28	+15 .99	+ 5 .88	-1′.50″.6
♄	— 0 .14			
	—20″.93	+15 .99	23″.11	-1′.50″.6

c. The eccentricity and the greatest equation of Jupiter, are increasing, from the action of Saturn; and those of Saturn decreasing, from the re-action of Jupiter. Their secular variations are nearly in the proportion of 1 to 2.

277. *Uranus*

277. *Uranus*, on account of his great distance, suffers no disturbance in his motion but from Saturn and Jupiter. The principal inequality depends on Saturn; which, if S be the longitude of that planet, U of Uranus, and A of the aphelion of Saturn, is $2'.30'' \sin(S-2U+A)$.

For the other equations, see LA LANDE, *Astron.* III. § 3671. Also the *New Tables*, VINCE, *Astron.* III. p. 106.

The orbit of Uranus is also disturbed. The node moves backward at the rate 34″.25 annually, and the aphelion forward at that of 2″.55. The eccentricity increases, and the secular variation of the greatest equation of the centre is + 11″.03.

When, in the preceding Tables, a planet is represented as producing a change in the place of its node, it must be understood that it does not produce this effect by its action on its own orbit, but by its action on the plane of the ecliptic.

278. *Comets*, in describing their elliptic orbits round the Sun, have been found to be disturbed by the action of the larger planets, Jupiter and Saturn; but the great eccentricity of their orbits, makes it impossible, in the present state of mathematical science, to assign the quantity of that disturbance for an indefinite number of revolutions, though it may be done

for

for a limited portion of time, by considering the orbit as an ellipsis, the elements of which are continually changing.

This is the method of La Grange, and is followed in the *Mecanique Celeste*, Part II. Chap. 9.

Dr Halley, when he predicted the return of the comet of 1682, took into consideration the action of Jupiter, and concluded that it would increase the periodic time of the Comet a little more than a year; he therefore fixed the time of the re-appearance to the end of the year 1758, or the beginning of 1759.

He professed, however, to have made this calculation hastily, or, as he expresses it, *levi calamo. Synopsis of the Astronomy of Comets.*

279. Clairaut, on calculating with great care and labour the effects both of Jupiter and Saturn, found that the return of the Comet would be retarded 511 days by the former, and 100 by the latter; in consequence of which he foretold that its return to its perihelion would be on the 15th of April 1759.

He said at the same time, that he might be out a month in his calculation. The Comet actually reached its perihelion on the 13th of March, just 33 days earlier than was predicted; thus affording

a

a very remarkable verification of the theory of Gravity, and the calculation of Disturbing Forces.

This Comet may be expected again about the year 1835. The investigations of LA PLACE will render it much easier to calculate the quantity by which its arrival may be anticipated or retarded by the action of the planets.

A Comet, which was observed in 1770, had a motion, when carefully examined, which could not be reconciled with a parabolic orbit, but which might be represented by an elliptic orbit of moderate eccentricity, in which it revolved in the space of five years and eight months. This Comet, however, had never been seen in any former revolution, nor has it been seen in any subsequent one.

280. Mr BURKHARDT, on tracing the path of this Comet, found that between the years 1767 and 1770, it had been very near to Jupiter, and again had come very near to that planet in 1779; he therefore conjectured, that the disturbance of Jupiter might have so altered its original orbit, as to render the Comet for a time visible from the Earth; and may have so changed it again, after one revolution, as to restore the Comet to the same region in which it had formerly moved. This conjecture has been confirmed by a careful application of the formulas of the *Mecanique Celeste.*

Mr

Mr BURKHARDT found that the Comet had come so near to Jupiter between 1767 and 1770, that it may have been brought from an orbit of which the semi-transverse was 13.293, (that of the Earth's orbit being 1), and in which it revolved in a period of 48.466 years, to one in which the semi-transverse was 3.178, and in which it revolved in five years and eight months, as it was at that time observed to do. While revolving in this orbit, it came near to Jupiter again; and its time of revolution and its distance were so changed, that the latter became 6.388, and the former 16 years. In this orbit it cannot, any more than in its first, come so near the Earth as to be visible.

The preceding is the greatest instance of disturbance that has yet been discovered among the bodies of our system, and furnishes a very happy and unexpected application of the theory of Gravitation.

281. Though the Comets are disturbed in so great a degree by the action of the Planets, they do not appear by their re-action to produce any sensible effects.

This must no doubt arise from the small quantity of matter which a Comet contains.

The Comet of 1770 came so near to the Earth, as to have its periodic time increased by 2.046 days, according to LA PLACE's computation, and if it had been equal in mass to the Earth, it would have made

made an augmentation of 2h 48m nearly in the length of the year. But it is certain, the same author adds, that if an augmentation so great as even two seconds in the length of the year, had taken place at that time, it would have been discovered from Dr MASKELYNE'S observations, when compared by DE LAMBRE for the construction of the *New Astronomical Tables.* The mass of the Comet, therefore, cannot have been $\frac{1}{500}$th of the mass of the Earth.

The same Comet passed through the midst of the satellites of Jupiter without producing the smallest effect. *Mecanique Celeste,* vol. IV. p. 250.

It is reasonable, therefore, to think, that no material, or even sensible alteration has ever been produced in our system by the action of a Comet.

SECT.

SECT. IV.

OF THE DISTURBANCES WHICH THE SATELLITES OF JUPITER SUFFER FROM THE ACTION OF ONE ANOTHER.

282. THE application of the same principles to the satellites of Jupiter, has fully explained all the irregularities which had been observed in their motions, and has reduced under known laws several others, of which the existence had been indistinctly perceived.

A very remarkable relation takes place between the mean motions of the first three satellites, as remarked § 185; the mean motion of the first satellite + twice that of the third, being equal to three times the mean motion of the second, reckoning from any instant of time. LA PLACE has shewn, *Mecanique Celeste*, liv. II. chap. 8. that if the primitive mean motion of these satellites was near this proportion, their mutual action on one another must in time have brought about an accurate conformity to it.

It

It follows, that Long. 1st Sat — 3 Long. 2d Sat. + 2 Long. 3d Sat. = a constant quantity; and it has been found, since ever the satellites were observed, that this constant quantity has been nearly equal to 180°. This last must be the result of original constitution.

283. The first satellite moves nearly in the plane of Jupiter's equator, and has no eccentricity except what is communicated to it from the third and fourth; the irregularities of one of these small planets producing similar irregularities in the rest. It has beside an inequality chiefly produced by the action of the second, and circumscribed by the period of 437.659 days.

284. The orbit of the second satellite moves on a fixed plane, to which it is inclined at an angle of 27′.13″, and on which its nodes have a retrograde motion, so that they complete a revolution in 29.914 years.

The motion of the nodes of this satellite is one of the principal *data* that have been used for determining the masses of the satellites, which are so necessary to be known in computing their disturbances. This satellite has no eccentricity but what it derives from the action of the third and fourth.

285. The

285. The third satellite moves on a fixed plane that is between the equator and the orbit of Jupiter, and is inclined to that plane at an angle of 12′.20″, its nodes making a tropical revolution, (retrograde), in 141.739 years.

The equator of Jupiter is inclined to the plane of his orbit, at an angle of 3°.5′.27″; the fixed planes on which the planes of the orbits move, are determined by theory, and probably could never be discovered by observation alone.

286. The orbit of the third satellite is eccentric; but appears to have two distinct equations of the centre; one which really arises from its own eccentricity; and another, which theory shews to be an *emanation* from the equation of the centre of the fourth satellite. The first equation is referable to an apsis, which has an annual motion of 2° 36′ 39″ forward in respect of the fixed stars; the 2d equation is referable to the apsides of the 4th satellite.

These two equations may be considered as forming one equation of the centre, referable to an apsis that has an irregular motion. The two equations coincided in 1682, and the sum of their *maxima* was 13′ 16″. In 1777, the equations were opposed, and their difference was 5′ 6″.

Observation

Observation alone led M. WARGENTIN to the knowledge of these inequalities, but he could not discover the law.

287. The orbit of the fourth satellite moves on a fixed plane, to which it is inclined at an angle of 14′ 58″, and its nodes complete a sidereal revolution (backward) in 531 years. The fixed plane itself is inclined at an angle of 24′ 33″ to the equator of Jupiter; the orbit is very sensibly elliptical, and the line of the apsides has an annual motion of 43′ 34″.7.

The motion of the apsides of this satellite is one of the principal *data* from which the quantities of matter have been determined.

288. If the mass of Jupiter be supposed unity, the mass of the 1st Sat. = .0000173281
of the 2d = .0000232355
of the 3d = .0000884972
of the 4th = .0000426591

If the mass of the Earth be put = 1, that of the third satellite will be found .027337. Now the mass of the Moon is $\frac{1}{68.5}$ = .014599, and therefore the quantity of matter in the third satellite is about twice as great as that in the Moon. The fourth satellite

satellite is nearly equal to the Moon. *Mecanique Celeste*, 2de Partie, liv. 8. chap. 8.

CONCLUSION.

289. One general result of these investigations is, that both in the system of primary and secondary planets, two elements of every orbit remain secure against all disturbance; the *mean distance*, and the *mean motion*, or, which is the same, the transverse axis of the orbit, and the time of the planet's revolution. Another result is, that all the inequalities in the planetary motions are periodical, and observe such laws that each of them, after a certain time, runs through the same series of changes.

Every inequality is expressed by terms of the form $A \sin nt$ or $A \cos nt$; where A is a constant coefficient, and n a certain multiplier of t the time, so that nt is an arch of a circle, which increases proportionally to the time. Now, though nt is thus capable of indefinite increase, yet since $\sin nt$ never can exceed the radius or 1, the inequality can never exceed A. Accordingly, the value of the term $A \sin nt$, first increases from 0 to A, and then decreases from A to 0; after which it becomes negative,

gative, extends to = — A, and passes from thence to 0 again; the period of all those changes depending on n the multiplier of t.

If into the value of any of the inequalities, a term of the form, $A \tan nt$, $\frac{A}{\sin nt}$, or of the form Ant, were to enter, the inequality so expressed, would continually increase, and the order of the system might finally be displayed.

La Grange and La Place, in demonstrating that no such terms as these last can enter into the expression of the disturbances of the planets, made known one of the most important truths in physical astronomy. They proved, that the planetary system is stable; that it does not involve any principle of destruction in itself, but is calculated to endure for ever, unless the action of an external power be introduced.

290. This accurate compensation of the inequalities of the planetary motions, depends on three conditions, belonging to the primitive and original constitution of the system:

I. That the eccentricities of the orbits are all inconsiderable, or contained within very narrow limits:

II. That the planets all move in the same direction, as both primary and secondary do from west to east:

III. That the planes of their orbits are but little inclined to one another.

But for these three conditions, terms of the kind mentioned above would come into the expression of the inequalities, which might therefore increase without limit.

These three conditions do not *necessarily* arise out of the nature of motion or of gravitation, or from the action of any physical cause with which we are acquainted. Neither can they be considered as arising from chance; for the probability is almost infinite to one, that, without a cause particularly directed to that object, such a conformity could not have arisen in the motions of thirty-one different bodies scattered over such a vast extent.

The only explanation, therefore, which remains, is, *that all this is the work of intelligence and design, directing the original constitution of the system, and impressing such motions on the parts as were calculated to give stability to the whole.*

SECT.

SECT. V.

OF THE ATTRACTION OF SPHERES AND SPHEROIDS.

291. THE particles of matter being supposed to attract with forces inversely as the squares of the distances, the forces with which similar solids attract particles similarly situated with respect to them, are as any two homologous lines, or any two similarly situated lines, in those solids.

This proposition is easily proved of pyramids, and *frusta* of pyramids, of which the solid angle is indefinitely small. From thence it is transferred to all solids which can be resolved into the same number of similar pyramids, that is, to all similar solids. MACLAURIN'S *Fluxions*, § 629.

292. The same law of attraction being supposed, a particle placed any where within a hollow shell of any thickness, contained between similar and similarly situated surfaces, will be *in equilibrio*, or will be urged equally in all opposite directions.

This

This follows readily from the preceding. The chief application of the proposition is to spherical or spheroidal crusts.

A spheroid is a solid bounded by a surface generated by the revolution of an ellipsis about either axis.

The matter of which the solids consist, is supposed in these propositions to be homogeneous.

293. A particle placed any where without a sphere, of which the parts attract, as in the last article, will be urged to the centre of the sphere, with a force that is as the quantity of matter in the sphere, divided by the square of the distance from its centre.

Princip. Math. lib. I. prop. 74. SIMPSON'S *Fluxions*, § 380.

If m be the mass of the sphere, x the distance of a particle from the centre, the attraction is $\frac{m}{x^2}$.

If the radius of the sphere $= r$, and the density of the matter contained in it $= d$, the quantity of matter, or $m = \frac{4 \pi r^3 d}{3}$, and so the attraction

$$= \frac{4 \pi r^3 d}{3 x^2}.$$

When

When $x = r$, this expression becomes $\frac{4 \pi r d}{3}$, which is therefore the attraction at the surface of the sphere.

The force of attraction or of gravity at the surfaces of different spheres, are therefore as their densities multiplied into their radii.

294. The force with which a particle placed any where within a sphere, is urged toward the centre of the sphere, is proportional to its distance from that centre.

From what goes before, it is plain, that it is not attracted by any part of the sphere more distant from the centre than itself. It is therefore attracted only by the sphere at the surface of which it is placed.

295. If a particle be placed any where in the interior of an elliptic spheroid; and if through it there be described a spheroidal surface, similar and similarly situated to the surface of the spheroid, the particle will be urged only by the attraction of the spheroid contained within this surface, and the force of that attraction will be the force acting on a particle similarly situated

on

on the surface of the exterior spheroid, as the transverse axis of the interior to the transverse axis of the exterior spheroid.

296. If in an oblate spheroid differing little from a sphere, b be the polar semi-axis, $b+c$ the radius of the equator, and φ the angle which a line drawn from the centre to a given point on the surface makes with the axis of the spheroid, f the force with which the given point is attracted by the spheroid;

$$f = \frac{4\pi b}{3}\left(1 + \frac{c}{b} \cdot \frac{4 - \sin^2 \varphi}{5}\right).$$

At the equator, where $\varphi = 90°$, this becomes $\frac{4\pi b}{3}\left(1 + \frac{c}{b} \times \frac{3}{5}\right)$. At the pole, where $\varphi = 0$,

it is $\frac{4\pi b}{3}\left(1 + \frac{c}{b} \times \frac{4}{5}\right)$.

For a very elementary demonstration of this theorem, see the Notes on NEWTON'S *Principia*, by MADAME DU CHASTELLET, *Principes*, &c. tom. II. p. 237.

297. If in an oblate spheroid, a be the semi-transverse axis, and e the eccentricity of the meridian; then, if gravity at the surface, in the plane

plane of the equator, be called g, the force of gravity at any distance x, from the centre, in the same plane, will be

$$\frac{g \times a^2}{x^2} \times \frac{1 + \frac{3 e^2}{10 x^2} + \frac{9 e^4}{56 x^4}.}{1 + \frac{3 e^2}{10 a^2} + \frac{9 e^4}{56 a^4}, \text{ \&c.}}$$

Maclaurin's *Fluxions*, § 659.

Supposing e, as in the former articles, to be the difference of the two semi-axes of the meridian, $e^2 = 2\,a\,c$, and the above will become,

$$\frac{g \cdot a^2}{x^2} \times \frac{1 + \frac{3\,a\,c}{5\,x^2}, \text{ \&c.}}{1 + \frac{3\,c}{5\,a}, \text{ \&c.}}$$

It is evident from this, that the attraction of a spheroid, in the plane of its equator, does not decrease exactly in the inverse ratio of the square of the distance; on which account a small inequality is produced in the motion of the satellites belonging to planets which are not entirely spherical. The Moon is subject to an inequality arising from this cause, as are also the satellites of Jupiter.

SECT.

SECT. VI.

FIGURE OF THE EARTH.

298. FROM observation it has already been inferred that the Figure of the Earth is nearly that of an oblate spheroid, of which the greater axis, the diameter of the equator, is to the less, the axis of revolution, as 312 to 311.

The strict meaning of the phrase, *the Figure of the Earth*, has already been defined, and must be carefully kept in view, in searching into the causes which have determined it.

Since the Earth revolves on its axis, it is evident, that its parts are all under the influence of a centrifugal force, proportional to their distances from that axis, and that if the mass were fluid, the columns toward the equator, being composed of parts that are lighter, must extend in length, in order to balance the columns in the direction of the axis. By this means an oblateness or elevation at the equator would be produced, similar, in some degree at least, to that which the Earth has been found to possess. Though it is not evident how the centrifugal force would produce such an effect

on

on a solid body like the Earth, it may throw some light on the matter, to inquire into the figure which a homogeneous fluid would put on, its parts being all supposed to gravitate to one another, and at the same time to be under the influence of a centrifugal force, by which they all tended to recede from an axis given in position.

To the equilibrium of a fluid mass, it is necessary that any two columns reaching to the surface from any point in the interior of the fluid, should balance one another, or should press equally on that point. Conversely, when this is the case, however many the forces may be that act on the particles of the fluid, the mass must remain *in equilibrio.*

It follows from this, when a fluid is at rest its surface is at every point perpendicular to the direction of the diagonal resulting from all the forces which act at that point.

299. This equilibrium of the columns will take place in a mass of homogeneous fluid revolving on an axis, if it be formed into an oblate spheroid, such that the polar semi-axis is to the radius of the equator, as the attraction at the equator diminished by the centrifugal force at the same place, to the attraction at the pole.

This proposition was first demonstrated by MACLAURIN. *Fluxions*, tom. II. § 636 to 641.

300. Hence

300. Hence the fraction which expresses the ellipticity of the meridian, or the excess of the equatorial above the polar radius, is to that which expresses the ratio of the centrifugal force at the equator, to gravity at the same place, as 5 to 4.

This is also demonstrated by MACLAURIN, and follows readily, from comparing the last theorem with those already stated concerning the attraction of spheroids.

301. The centrifugal force is to the gravity at the equator, nearly as 1 to 289.

A body under the equator falls in a second through 15.0515 French feet, as is concluded from the length of the pendulum vibrating seconds.

Now the centrifugal force is measured by the deflection of a body from the tangent produced by the Earth's rotation in 1 second. But a point of the equator in a second of mean solar time, describes an arch of 15″.0417 of a degree; the versed sine of which, multiplied into the radius of the equator, in French feet, gives the deflection, which is found to be $\frac{1}{289}$ of the direct descent.

302. A

302. A homogeneous fluid, therefore, of the same mean density with the Earth, and revolving on an axis in the space of 23^h 56^m 4^{sec} of solar time, would be *in equilibrio* if it had the figure of an oblate spheroid, of which the axis was to the equatorial diameter as 230 to 229.

This is accordingly the figure which NEWTON ascribes to the Earth. *Princip.* lib. 3. prop. 19.

In the spheroid thus constituted, the gravitation at the equator will be to that at the pole as 229 to 230; and if g be the gravitation at the equator, and g' that at any other point, of which the latitude is λ, $g' = g\left(1 + \frac{\sin^2 \lambda}{230}\right)$; also if l be the length of the pendulum that vibrates seconds at the equator, and l' the length of a pendulum that does the same in Latitude λ, $l' = l\left(1 + \frac{\sin^2 \lambda}{230}\right)$.

NEWTON's investigation of the figure of the Earth, though very ingenious, involved some assumptions which prevented it from being quite satisfactory. A very accurate and elegant demonstration was afterwards given by MACLAURIN; and the investigation was improved and rendered more analytical by CLAIRAUT, *Fig. de la Terre.*

But

But though it was thus demonstrated that the parts of a homogeneous fluid, on which the figure of the oblate spheroid just described was any how induced, would be *in equilibrio*, yet it was not shewn conversely, that, whenever an equilibrium takes place in such a fluid mass, the figure of the mass must be the oblate spheroid in question. D'ALEMBERT indeed shewed, that there are more spheroids than one in which the state of equilibrium may be maintained; and this result, though it was not observed by MACLAURIN, might have been inferred from his solution. LE GENDRE afterwards proved, that the solids of equilibrium must always be elliptic spheroids, and that in general there are two spheroids which satisfy the conditions.

In the case of a homogeneous mass of the mean density of the Earth, revolving in the space of $23^h\ 56'\ 4''$, one of the spheroids is that which has been mentioned; the other, is one in which the equatorial diameter is to the polar, as 681 to 1. *Mém. Acad. des Sciences*, 1784. LA PLACE has added the limitation which follows.

303. A fluid and homogeneous mass, of the mean density of the Earth, cannot be in equilibrium with an elliptic figure, if the time of its rotation be less than $2^h\ 25^m\ 17^{sec}$; if the time of revolution is greater than this, there will always

ways be two elliptic spheroids, and not more, in which an equilibrium may be maintained.

If the density of the fluid is greater than the mean density of the Earth, the time of rotation with which the equilibrium ceases to be possible, is had, by dividing $2^{h}\ 25^{m}\ 17^{sec}$ by the square root of the density of the fluid, that of the Earth being unity. LA PLACE, *Theorie du Mouvement et de la Figure des Planetes*, Paris, 1784, p. 126.

304. If the fluid mass, supposed to revolve on its axis, be not homogeneous, but be composed of strata that increase in density toward the centre; the solid of equilibrium will still be an elliptic spheroid, but of less oblateness than if it were homogeneous.

This was demonstrated by CLAIRAUT, *Theorie*, &c. NEWTON fell into the mistake of supposing the contrary to be the case, or that the greater density toward the centre would be accompanied with greater oblateness. If the density increase, so as at the centre to be infinite, the ellipticity is $\frac{1}{2 \times 289} = \frac{1}{578}$, which is the case of the least ellipticity. $\frac{1}{230}$ is the case of the greatest.

305. Hence

305. Hence, as the ellipticity of the Earth has been shewn to be less than $\frac{1}{230}$, (viz. $\frac{1}{312}$ nearly), it is evident, that if the Earth is a spheroid of equilibrium, it is denser toward the interior.

The greater density of the Earth toward the centre, is in itself probable, and has been put beyond all doubt by very accurate experiments, made on the sides of the mountain Schehallien in Perthshire, by the late Dr MASKELYNE.

By observations of the zenith distances of stars, the difference of the latitude of two stations on the south and north sides of that mountain, was determined. A trigonometrical survey of the mountain also ascertained the distance between the same two points, and thence, from the known length of a degree of the meridian, under that parallel, the difference of the latitude of the stations was inferred, and was found less by 11″.6 than by the astronomical observations.

The zeniths of the stations had therefore been separated from one another more than in the usual proportion of the meridian distance; and this could only arise from the plummet on each side being attracted toward the body of the mountain.

From

From the quantity of this change of direction, the ratio of the attraction of the mountain to the attraction of the whole Earth, or to the force of gravity, was found to be that of 1 to 17804.

The bulk and figure of the mountain being also given from the survey, the mean density of the mountain was found to be to the mean density of the Earth nearly as 5 to 9. *Phil. Trans.* vol. LXVIII. p. 781. Also HUTTON's *Tracts*, vol. III. p. 62.

The mean density of the Earth, is therefore nearly double the density of the rocks which compose Schehallien, which seem again to be considerably more dense than the mean of those which form the exterior crust of the Earth. *Phil. Trans.* 1811, p. 347.

306. If the density of the Earth increase toward the centre, the ellipticity of the spheroid, and the diminution of gravity from the pole to the equator will not both be expressed by the same fraction as in the case of the homogeneous spheroid, but the sum of the two fractions will always be equal to the same quantity, viz. $\frac{1}{115}$ or .008695.

This theorem was first given by CLAIRAUT, and is of great use, as serving to explain the connection between the ellipticity, as ascertained from the measurement

surement of degrees, and from experiments with the pendulum.

As, in the actual figure of the Earth, the compression, or the ellipticity, is nearly .0032, § 60. if we take this from .008695, the remainder, .005495, or $\frac{1}{182}$, is the diminution of gravity from the pole to the equator. And the gravitation at any other point of the spheroid, is $g\ (1 + .005495 \sin^2 \lambda)$, g being the gravity at the equator. The length of an isocronous pendulum is expressed by the same formula. This agrees nearly with the observations on the length of the pendulum in different latitudes. See a Table of them, VINCE, *Astron.* vol. II. p. 105. LA LANDE, *Astron.* § 2712. BIOT, *Astron.* tom. III. p. 148.

The lengths of the pendulum in different latitudes, are less subject to irregularities than the lengths of degrees; the intensity of gravity being, as might be expected, less affected by local variations than its direction.

307. The inequalities on the surface of the Earth, and the unequal distribution of the rocks which compose it, with respect to density, must produce great local irregularities in the direction of the plumb-line, and are probably the causes of the inequalities observed in the measurement

surement of contiguous arches of the meridian, even where the work has been conducted with the greatest skill and accuracy.

This is exemplified in the great arch of the meridian measured across France; and in those measured in England and Hindostan. The cause may be something concealed under the surface, which can at present only be a subject of hypothetical, or, at best, of analogical reasoning.

These irregularities are so considerable, that the spheroid which agrees best with the degrees in France, is one having an ellipticity of $\frac{1}{152}$, nearly double of what may be accounted the mean ellipticity.

308. Notwithstanding these irregularities, the figure of the Earth has made such an approximation to the spheroid of *equilibrium*, as indicates either the original fluidity of the entire mass, or the gradual acquisition of a spheroidal figure in consequence of the repeated waste and reconsolidation of the parts near the surface.

If the whole mass of the Earth was ever in a fluid state, it must have been so from the action of heat. The insolubility of the greatest part of rocks and minerals in water, and the immense bulk of water that would be required for dissolving even those that are soluble, are insuperable objections to the hypothesis of aqueous formation. The igneous formation is not subject to either of these difficulties.

The spheroidal figure may also have been gradually acquired without supposing the original fluidity of the whole mass.

309. If in a terraqueous body, however irregular in its primitive form, the prominent parts are subject to be worn down, and the *detritus* to be carried to the lower parts, occupied by water, where they acquire a horizontal stratification, and are, by certain mineral operations, afterwards consolidated into stone; such a body, in the course of ages, must acquire a surface every where at right angles to the direction of gravity, and consequently more or less approximating to a spheroid of equilibrium.

The natural history of the Earth gives great countenance to the suppositions here introduced; which therefore seem to furnish a very rational explanation of the ellipticity belonging to the Earth

Earth, and to the planets which are known to revolve on their axes.

The distribution of the solid materials in the interior of the Earth, will very much affect the nature of this solid, and the manner in which the figure is acquired must probably prevent the approximation from ever being complete.

The distribution of the materials at any distance under the surface is unknown, and we have no means of examining into it, but by the measurement of degrees, or by experiments on the pendulum, and on the plumb-line, like those at *Schehallien*.

310. On the whole, the facts known, from observation, about the figure of the Earth, agree in general with theory; but there are in the expressions of that theory, so many quantities which are yet indeterminate, that the perfect coincidence of the two cannot be affirmed.

SECT.

SECT. VII.

OF THE PRECESSION OF EQUINOXES.

311. THE precession of the equinoxes, is the slow angular motion by which the intersection of the equator and ecliptic goes backward, at the rate of $50''\frac{1}{3}$ annually; while the inclination of these planes continues nearly the same; so that the pole of the equator describes a circle about the pole of the ecliptic in the space of 25748 years nearly.

In seeking for the cause of this phenomenon, it is natural to inquire how the gravitation toward distant bodies, such as the Sun and Moon, may affect the Earth's rotation on its axis.

312. From what has been proved of the force with which the Sun disturbs the motion of the Moon, it is evident, that every particle in that hemisphere of the Earth which is turned toward the Sun, is drawn toward that body, while every particle in the other hemisphere is drawn in the opposite direction, the force that acts

on

on any particle being as its distance from the plane that separates these hemispheres.

From this it follows, that if the Earth were a perfect sphere, the solar forces acting on the opposite hemispheres, would exactly balance one another, and could produce no motion in the Earth or its axis.

313. The Earth may be considered as a sphere circumscribed by a spheroidal shell or *meniscus*, thickest at the equator. The tendency of the Sun's action on this meniscus, except at the time of the equinoxes, is always to make it turn round the intersection of the equator with the ecliptic, towards the plane of this latter circle.

For the matter of the meniscus may be regarded as forming a ring round the Earth, in the plane of the equator. Now, the solar force acting on the part of this ring that is above the ecliptic, may at every point be resolved into two; one of which is in the plane of the equator, and the other perpendicular to it. The result of all the latter, must be a force tending to impress on the ring a motion round its intersection with the ecliptic. The same holds of the half of the ring that is under the ecliptic.

314. Hence, if the equator had no other motion, it would turn round its intersection with the

the ecliptic, till it coincided with that plane; the line of their intersection remaining all the while at rest.

The mean quantity of the solar force which is thus decomposed, is $\frac{3\,m}{a^3}\,r \cos\delta$, where m is the mass of the Sun, a the mean distance of the Sun from the Earth, r the radius of the equator, and δ the declination of the Sun.

The part of this force which is perpendicular to the plane of the equator, and which tends to make it move round the line of its nodes, is,

$$\frac{3\,m}{a^3}\,r \cos\delta \times \sin\delta.$$

315. As the ring which surrounds the equator, at the same time that it has the tendency just described, revolves on an axis perpendicular to its plane in twenty-four hours, it will not revolve on either of these axes, but on one in the same plane which divides the angle between them, so that the sine of its angular distance from each axis, is in the inverse ratio of the angular velocity round that axis.

If the arch, round the intersection of the equator and ecliptic, which the solar force acting upon the ring of the *meniscus*, would make the Earth describe

scribe in $\dot{t}$ an indefinitely small portion of time, be φ; and if the angle described in the same time by the diurnal revolution be v, the axis of that revolution will be changed by an angle of which the tangent is $\frac{\varphi}{v}$.

The substitution of the tangent for the ratio of the two sines mentioned in the proposition, is made, because the angle which the axis of the diurnal revolution makes with the intersection of the equator and ecliptic, is a right angle.

316. If the force which produces the motion φ, were, like that which produces the motion v, a force that acts only once, and leaves its effect to continue without diminution or increase, the change thus made in the position of the Earth's axis would be permanent, and no farther variation would take place; but the force that produces φ being one that acts continually, with more or less intensity, the quantity φ is constantly renewed, and the position of the Earth's axis is subject to continual change.

The relation between the momentary precession and φ, the momentary change in the position of the axis, will be made plain by a diagram.

Let

Let ABC (fig. 27.) be the ecliptic; the order of the letters A, B, C, marking the order of the signs; so that if A be the vernal, B is the autumnal equinox; AGB a circle perpendicular to the ecliptic; G the pole of the ecliptic; P the pole of the equator; and APB the equinoxial colure. HG=GK= the obliquity of the ecliptic; HPK a parallel to the ecliptic passing through the pole of the equator.

The equator is not represented in this figure; but the half of it, on the side turned toward the eye, is supposed to be under the ecliptic; its plane being at right angles to PP.

The change which the motion φ makes on the pole P, tends to bring it nearer to A. For the diurnal revolution about the axis PD, being in the direction ABC, and the motion about the axis AB being in the direction PG′, it will be in the quadrant PA that these motions will be opposed, so as at a certain point P′ (and in all the points of the line P′D) to destroy one another. PP′ is the arch, of which the tangent has already been shewn to be $\frac{\varphi}{v}$.

If a great circle be now described from the pole P′, it will be the new equator, and the point in which it intersects the ecliptic, will be the new equinox. The relation of these small variations to one another,

another, will be best seen by projecting a part of the circles on a plane touching the Earth in A.

If AC (fig. 27. N° 2.) be a small part of the ecliptic represented by a straight line; AP′ a like portion of the solstitial colure; and AQ of the equator. Then if AA′ be taken equal to PP′, in the former figure; and if A′C be drawn perpendicular to A′P′, it will represent the new equator; and AC, which is evidently *in antecedentia*, will represent the precession, and will be $= PP' \times \frac{1}{\text{sin obl.}} = \frac{\varphi}{v \text{ sin obl.}}$.

The Earth, however, cannot permanently revolve about any diameter that is oblique to the plane of the ring, because the centrifugal forces on opposite sides of such a diameter would not balance one another, and must therefore tend to bring the plane of the ring to be perpendicular to the new axis of rotation, or to make the primitive axis of the diurnal rotation coincide with this last.

The whole system of change, therefore, will begin anew, as in the former case, and the pole of the equator will constantly move *in antecedentia* round G. But the curve in which it will move is the circle HPK; for as the great circle APB touches this last in P, and as the arch PP′ is very small, P′ will deviate from the circumference of the latter circle, only by a quantity that is evanescent in respect of PP′. P′ will therefore describe a circle round

round G, moving *in antecedentia*, at the same rate with the line of the equinoxes.

317. If, therefore, the equator had received no motion but from the solar force, its inclination to the ecliptic would have changed continually; but the line of its intersection with that circle would have remained at rest. In consequence of the diurnal revolution, this effect is entirely reversed; the inclination of the two planes remains constant, and their intersection continually revolves.

This paradox is remarked by LA PLACE, *Systeme du Monde*, p. 270. His solution of it is not so elementary that it could be adopted here.

318. In order to calculate the precise measures of the different actions that have now been traced, the momentum with which the solar force would make the spheroidal *meniscus* already referred to, begin to revolve about the line of the equinoxes, supposing it to have no other motion, must be determined; the whole of the terrestrial spheroid must tend to revolve about the said axis with a momentum just equal to this quantity.

NEWTON, who first resolved the problem of the precession, was in an error with respect to this part of of the investigation, supposing that it was the quantity of *motion* (not of *momentum*) that must be equal in the meniscus and in the entire spheroid. D'ALEMBERT was the first who corrected this mistake.

When the angular motion is computed, t being the time of the diurnal, T of the annual revolution of the Earth, c the compression at the poles, and a the radius of the equator, δ the declination of the Sun; it is found that $\frac{\phi}{v} = \frac{3t^2}{T^2} \times \frac{c}{a} \sin\delta \cos\delta$.

This is the same with SIMPSON'S Theorem, *Miscellaneous Tracts*, London, 1757, Prob. 2. p. 14.

Next, supposing t and δ to vary, so as to get the fluxion of the precession, and afterwards integrating, the precession for an entire year is found

$= 360 \times \frac{3t}{2T} \times \frac{c}{a} \cos \text{Obl.}$ SIMPSON, *ibid.* p. 17.

From this, SIMPSON computes the precession arising from the Sun's action, supposing the Earth to be a homogeneous spheroid, and its compression $\frac{1}{231}$ to be 21″.1. Professor ROBERTSON of Oxford makes the precession 21″.03. *Phil. Trans.* 1807. p. 82. FRISIUS makes it 21″.$\frac{1}{4}$ on the same hypothesis. *Theoria Diurni Motûs:* Opera, tom. III. p. 288.

If

In these computations, the compression is $\frac{1}{231}$. We have seen that its true value is $\frac{1}{312}$. The quantity of the precession diminished in this ratio is $15''.3$.

319. The Moon produces a similar retrogradation in the intersection of the equator with the plane of the lunar orbit, which, being proportional to the cosine of the inclination of these planes, is subject to continual variation; but the mean quantity of the precession which is thus produced, relatively to the ecliptic, is the same as if the Moon moved in the plane of that circle.

LA PLACE, *Systeme du Monde*, liv. 4eme, chap. 13. p. 271. 2de edit.

If the precession due to the Sun's force be $15''.3$, that which is produced by the Moon is $35''$; which is to the former nearly as 7 to 3. If the effect of the Sun were reduced to 12.5, that of the Moon would be triple of it, which is agreeable to the latest results deduced from the theory of the tides, as will be seen in the next section.

320. The

320. The action of the Moon produces also an inequality which diminishes the precession by a quantity proportional to the sign of the distance of the Moon's ascending node from the vernal equinox ; and, besides this, a diminution in the obliquity of the ecliptic, proportional to the cosine of the said distance.

These two inequalities constitute the Nutation ; and the result here stated from the theory of gravitation, is conformable to that which was before given from observation, § 218. Conformably to what is said there, these two inequalities may be expressed by the revolution of the extremity of the Earth's axis produced to the Heavens, and describing an ellipse, as there represented ; of which the greater axis is to the less, as the cosine of the obliquity of the ecliptic to the cosine of twice that obliquity.

321. The precession, on account of variations in the solar action, as well as in the lunar, is subject to some inequalities, not included in the preceding theorems.

The amount of the precession, including all these inequalities, may be calculated for any period of time, by a formula given in the *Mecanique Celeste*, which, reduced to the sexagesimal notation, is

$$50''.412 \times t - 4626'' \sin. 1''.394 \times t + 12954'' \sin 1''.4298 \times t.$$

t is the number of years reckoned from 1750.

As

As the annual precession is not always the same, the length of the tropical year, in remote ages, has been somewhat different from what it is at present. In the age of HIPPARCHUS, it was about 10″ longer. The siderial year, as already observed, remains invariable.

Diurnal Rotation.

322. The velocity of the Earth's rotation on its axis, or the length of the day, is not affected by the action of the Sun or Moon, in such a degree as can ever become sensible, even to the nicest observation.

Systeme du Monde, p. 271. The small inequalities so produced, do not accumulate by time, but quickly compensate one another.

323. The motions of bodies near the surface of the Earth, tend, in some cases, to alter the velocity of the diurnal rotation; but if these motions are only such as we at present perceive, their effects, like the preceding, must for ever remain insensible.

A body which, by descending from a height, or by moving from the equator toward the poles, comes nearer to the Earth's axis, tends to accelerate the motion

motion about that axis, because it brings with it, into its new situation, more velocity than it can retain, and must therefore communicate a part of it to the general mass. The contrary happens when a body recedes from the axis, and the quantity of the acceleration or retardation thus produced, may be calculated, on the principle, that the total momentum of the Earth must remain the same, notwithstanding of any change that can arise from the action of its parts on one another.

From this it follows, that if any body change its distance from the axis, the momentum of rotation of the whole Earth, is to the change in the momentum of rotation of the body, as the velocity of the diurnal rotation to the variation in that velocity, arising from the motion of the body.

As the first of these terms is incomparably greater than the second, so must the third be than the fourth.

Hence, though the degradation of mountains, and the carrying of matter by the rains from a higher to a lower level, are effects that go on continually, the amount can never be so great as to be sensible.

In some cases, even these small changes are immediately compensated. The constant current in the atmosphere, from the poles to the equator, is counteracted, in the retardation it tends to produce, by the contrary current, which, in the superior

rior regions of the air, sets from the equator toward the poles.

324. In former ages, however, if the changes that have happened on the surface, or in the interior of the Earth, have been as great as some Geologists suppose, the change in the diurnal rotation may have been very considerable.

If the ocean once stood at the height of 15,000 feet above its present level, a quantity equal in bulk to a 440th part of the whole Earth, must have passed from being above the level of the present sea to be under it. If the mean density of water were the same with that of the Earth, it may be calculated easily, that the time of the diurnal revolution must, on this account, be shortened by $5^{m}.682$. As the Earth's mean density is to that of water nearly as 4.71 to 1, this acceleration is reduced to 1^{m} 12^{sec}.

325. The changes on the surface, or in the interior of the Earth, may have produced great variations in the *position* of the Earth's axis, relatively to the Earth itself, as well as in the *time* of the diurnal rotation.

If the Earth had originally a very irregular figure; and if, as above suggested, it has acquired its present

sent form, by the wearing down of the more prominent parts, and their subsequent consolidation in the form of horizontal strata, the axis of the Earth's rotation may have been very different from what it now is; it may have gone through a long series of changes, and may have carried the equator, and the accumulation of waters which accompanied it, over regions from which they are now far distant.

Many facts in the natural history of the Earth, and of the mineral kingdom, give countenance to these suppositions; and if it be true that the more ancient strata have been set on edge, and that continents have been raised up by the action of an expansive force in the interior of the Earth, such action may have materially assisted in changing the position of the Earth's axis.

Obliquity of the Ecliptic.

325. The position of the ecliptic is subject to change by the action of the planets; each of them producing a retrograde motion in the intersection of the plane of the ecliptic with the plane of its own orbit. This does not affect the inclination of these two planes, nor does it affect the plane of the equator, but it neverthe-

 less

less changes the inclination of the ecliptic to the equator, and also the line of their intersection,

This change in the inclination, and in the position of the line of the equinoxes, is easily deduced by spherical trigonometry from the retrogradation of the intersection of the two planes, and from the constancy of their inclination. See LA LANDE, *Astron.* § 2751, &c.

326. The variations in the obliquity of the ecliptic, thus produced, are among the number of the secular inequalities which have long periods, and, after reaching a *maximum*, return in a contrary direction.

As far back as observation goes, the obliquity of the ecliptic has been diminishing, and is doing so at present, by 52″ in a century; it will not, however, always continue to diminish, but in the course of ages will again increase, oscillating backwards and forwards on each side of a mean, from which it never can depart far.

The secular variation of the obliquity was less in ancient times than it is at present; it is now near its *maximum*, and will begin to decrease in the 22d century of our era.

LA GRANGE has shewn, that the total change of the obliquity, reckoning from that in 1700, must be less than 5° 23′; *Mém. Acad. de Berlin*, 1782, p. 284.

p. 284. Also, that the changes in the inclinations of the planetary orbits, are all periodical, and cannot carry the planes of those orbits beyond the limits of the zodiac, or 8° on either side of the ecliptic. By the retrogradations of the nodes of the ecliptic and the planetary orbits, the precession of the equinoxes is diminished by a small quantity, which is at present about 0″.261 annually. *Ibid.* p. 281. All this is quite independent of the figure of the Earth, and would be the same though the Earth were truly spherical.

THE first solution of the problem of the Precession was given by NEWTON, *Princip.* lib. III. prop. 39. It is not free from error; but it displays, in a strong light, the resources of genius contending with the imperfections of a science not sufficiently advanced for so arduous an investigation. One mistake has already been pointed out, § 318; another consisted in supposing, that the motion of the nodes of the Ring surrounding the equator, must be the same with those of the Moon, allowing for the different time of rotation, and the different inclination to the ecliptic.

D'ALEMBERT corrected these mistakes, and gave an accurate, though prolix solution, which also comprehended in it the theory of the Nutation. *Precession*

des

des Equinoxes, Paris, 1749. A solution equally correct and original, was given about the same time by EULER, *Mém. Acad. de Berlin*, 1749.

Two solutions, in the *Philosophical Transactions* for 1754 and 1756, continued to follow the method of NEWTON. The first of these was by SYLVABELLE, the second by WALMESLEY; and this last, though it retained both the defects just mentioned, is remarkable for the elegance of the demonstrations. It extended the problem to the nutation of the Earth's axis, and it treated of the diminution of the obliquity of the ecliptic by the action of the planets.

A memoir by LA GRANGE, on the *Libration* of the Moon, which was crowned by the Academy of Sciences at Paris in 1769, contained an excellent solution of the problem of the Precession.

SIMPSON, in his *Miscellaneous Tracts*, has given the solution already referred to, which is one of great simplicity and correctness. Its only defect is, that it does not clearly point out the means by which the uniform inclination of the Earth's axis is maintained.

Another very elegant solution, is that of FRISI above referred to; *Theoria Diurni Motûs, Opera*, tom. III. p. 292.

LA LANDE has followed SIMPSON, as has also VINÇE, in his *Astronomy*. The latest solution is that of Professor ROBERTSON of Oxford, *Phil. Trans.* 1807;

1807 ; it is also after the method of SIMPSON, and the investigation is accurate and concise.

The solution of LA PLACE, in the *Mecanique Celeste*, must be considered as the most perfect, and that which can most certainly be said to include, and to estimate with accuracy, all the causes which have a share in this phenomenon. There is, however, one defect it may be said to have, that as it proceeds entirely by the calculus, it does not sufficiently carry the imagination along with it.

SECT. VIII.

OF THE TIDES.

327. THE alternate rise and fall of the surface of the sea twice in the course of a lunar day, or of 24^h 50^m 48^{sec} of mean solar time, is the phenomenon known by the name of the *Tides*.

The time from one high-water to the next, is, at a mean, 12^h 25^m 24^{sec}. The instant of low-water is nearly, but not exactly, in the middle of this interval; the tide, in general, taking nine or ten minutes more in ebbing than in flowing.

At the time of new and full Moon, the tide is the highest, and the interval between the consecutive tides is the least, viz. 12^h 19^m 28^{sec}. At the quadratures, or when the Sun and Moon are 90° distant, the tides are the least, and the interval between them is the greatest, viz. 12^h 30^m 7^{sec}.

328. The time of high-water is mostly regulated by the Moon, and in general, in the open sea, is from two to three hours after that planet has

has been on the meridian, either above or under the horizon.

On the shores of the larger continents, and where there are shallows and obstructions, there are great irregularities in this respect, and when these exceed six hours, it may seem as if the high-water preceded the Moon's transit over the meridian.

For any given place, the hour of high-water is always nearly at the same distance from that of the Moon's passage over the meridian.

This constant dependence on the Moon, is the reason why the tides are considered as astronomical phenomena.

329. Though the tides seem to be chiefly regulated by the Moon, they appear also, in a certain degree, to be under the influence of the Sun. Thus, at the syzygies, when the Sun and Moon come to the meridian together, the tides, *cæteris paribus*, are the highest; at the quadratures, or when the Sun and Moon are 90° distant, the tides are least. The former are called the *Spring*, the latter the *Neap Tides*.

The highest of the Spring Tides is not the tide that immediately follows the syzygy, but is in general the third, and, in some cases, the fourth.

At

At Brest, the tides of the syzygies rise to 19.317 feet; and those of the quadratures to 9.151, not quite the half of the former quantity.

In the Pacific Ocean, the rise, in the first case, is 5 feet; in the second, 2 or 2.5.

The greater the rise of high-water above the level of a fixed point, the greater the depression of the corresponding low-water relatively to the same point. To estimate the height of the tide, it seems best to take the excess of the mean of the two consecutive high-water marks, above the intermediate low-water. This is the method of LA PLACE.

330. The height of the tide is affected by the vicinity of the Moon to the Earth, and increases, *cæteris paribus*, when the parallax and apparent diameter of the Moon increase, but in a higher ratio.

The greatest variation of the Moon's semidiameter above or below the mean is about $\frac{1}{15}$th of the whole, and the corresponding variation of the tide at the syzygies is ^{3} th of its mean quantity. *Systeme du Monde*, p. 77.

331. The rise of the tide is affected by the declination of the luminaries; it is greatest, *cæteris paribus*, at the equinoxes, and least at the solstices.

When

When the Moon is in the northern signs, the tide of the day, in all northern latitudes, is somewhat greater than the tide of the night.

The contrary happens when the Moon is in the southern signs.

332. If the tides be considered relatively to the whole Earth, and to the open Sea, it is evident, that there is a meridian, about 30° eastward of the Moon, where it is always high-water, both in the hemisphere where the Moon is, and in the opposite; on the west side of this circle, the tide is flowing; on the east, it is ebbing; and on the meridian, at right angles to the same, it is everywhere low-water.

These meridians move westward, preserving nearly the same distance from the Moon, only approaching nearer to her at the syzygies, and going farther off at the quadratures.

In high latitudes, whether south or north, the rise and fall of the tides are inconsiderable. It is probable that at the poles there are no tides.

The great *Wave* which, in this manner, constitutes the tide, is to be considered as an undulation, or reciprocation of the waters of the ocean, in which there is, except when it passes over shallows, or ap-

proaches

proaches the shores, little or no progressive motion.

In all this, no regard is had to the operation of local causes, winds, currents, &c. by which these general laws are modified, or overruled.

333. The dependence of the phenomena just enumerated on the motion of the Sun and Moon, naturally suggests an inquiry into the effects which the action of these bodies may produce on the waters which cover so large a proportion of the Earth's surface.

334. If m be the mass, and a the distance of the Sun, r the mean radius of the Earth, z the distance of the Sun from the zenith of any place (or the distance of that place from the point to which the Sun is vertical) a particle of matter at that place is drawn toward the Sun by a force equal to $\frac{3\,m\,r}{a^3}\cos z$; besides having its gravity increased by another force $= \frac{m\,r}{a^3}$.

This is derived from the resolution of forces, in the same way as when the Sun's action on the Moon was investigated.

When

When the point is in the hemisphere opposite to the Sun, cos z becomes negative, so that the force draws the particle from the Earth, in a direction opposite to that which it has in the other hemisphere.

If m' be the mass of the Moon, a' its distance from the Earth, z' the distance of the Moon from the zenith, the forces by which the Moon affects the gravity of bodies on the Earth's surface, is expressed by the same formulas.

335. The preceding force may again be resolved into two, one of which, in the direction of the Earth's radius, and opposite to gravity, is $\frac{3\,m\,r}{a^3}\overline{\cos z}^2$, and, with that already mentioned, as increasing gravity, viz. $\frac{m\,r}{a^3}$, makes the whole force by which the Sun's action diminishes gravity, to be $\frac{3\,m\,r\cos z^2 - m\,r}{a^3}$.

The other force, draws the particles everywhere horizontally toward the point that has the Sun in its zenith, and is equal to $\frac{3\,m\,r}{2\,a^3}\sin 2\,z$. In the he-

misphere

misphere opposite to the Sun, this attraction is toward the point which has the Sun in its *nadir*.

336. If the surface of the Earth be covered by a fluid, the parts will move in obedience to these forces; those columns, where the Sun is vertical, will be rendered lighter, and will be lengthened, in order to be *in equilibrio* with those at the distance of 90°, which will of course be shortened.

The same thing exactly will happen in the opposite hemisphere; and if the waters on the Earth's surface were at rest, the Ocean would form itself into an oblong spheroid, with its longer axis passing through the attracting body. When there are two attracting bodies, their effects will most nearly coincide, when they are nearest to one another; and will be equal to the sum of the two effects taken separately. When they are 90° distant, the effect produced will be the difference of the separate effects. This is NEWTON's explanation of the tides. He calculates the mean force by which the Sun diminishes the force of gravity at the surface, as equal to one 12868200 part of gravity. Hence he deduces the longer axis, and the greatest rise of the tide produced by the Sun at 9.2 inches. *De Systemate Mundi. Opera*, tom. III. p. 212. A different determination is given in the *Principia*, lib. III. prop. 36.

The

The force drawing the water horizontally is not included here; its tendency is to increase the effect just calculated.

The force $\frac{mr}{a^3}$, by which the solar force everywhere increases gravity, need not be taken into account, and does not affect the equilibrium of the water.

The height to which the sea would thus rise at high-water, above the level which it would stand at if acted on by gravity alone, is twice as great as its depression under that level at low-water. This is easily demonstrated from the content of the spheroid remaining always the same.

337. The preceding is sufficient to shew, that the phenomena of the tides are effects that might be expected from the principle of gravitation. It is, however, an approximation from which exact results cannot be obtained, since a material element has been left out, namely, the motion of the water, on which the forces of the Sun and Moon are exerted.

338. The rapid motion of the waters, in consequence of the diurnal rotation, prevents them from assuming, at every instant, the figure which

which the equilibrium of the forces acting on them would require; so that they oscillate continually, alternately approaching to that figure, and receding from it.

To resolve the problem of the tides, including the condition of the diurnal motion, is a matter of great difficulty, and requires all the latest improvements of the *Calculus*. LA PLACE is the only one who has ventured to undertake it, and he calls it "*le Probleme le plus épineux de toute la Mecanique* "*Celeste.*" *Exposition du Syst. du Monde*, chap. 10. p. 248. edit. 2de. His solution agrees with the phenomena more nearly than could be expected, considering that the continuity of the Ocean is much interrupted by the land, and that this, together with the inequalities in the depth of the sea, the friction against the bottom and sides, &c. are all causes impossible to be submitted to analysis.

339. The oscillations excited in the Sea, by the force of the Sun, and by that of the Moon, are similar to one another, and are so combined, that each effect takes place in its full amount, independent of the other.

The law which connects the rise and fall of the tide with the time at any place, is expressed by a geometrical construction.

340. Let

340. Let AB, (fig. 28.) be the total rise of the tide at any place; and let the circle ADBF be described on it. If we suppose this circle placed vertically, with the point B at the level of low-water, the tide will rise or fall over equal arches, in equal times; that is, if the arches BD, DH, HA, be equal, the times in which the water will rise from B to D, from D to H, &c. will be equal; and the same in the ebbing of the tide.

Mecanique Celeste, tom. II. p. 221.

If the arch BD be proportional to the interval between any instant and the time of low-water, BE the versed sine of that arch, will be the height to which the water will have risen.

Near the high and low water, the differences of the depths from those of the high or low water, are as the squares of the times.

341. The total rise of the water in the tides of the syzygies, is to the total rise at the quadratures nearly as 2 to 1; and hence the effect of the Moon is also nearly to that of the Sun as 3 to 1.

This is known from observation, and hence the mass of the Moon, in respect of that of the Sun, and consequently in respect of the mass of the Earth, may be found.

For

For since $\frac{m}{a^3}$ and $\frac{m'}{a'^3}$ are as the forces of the Sun and Moon on the waters of the Sea, $\frac{m}{a^3} : \frac{m'}{a'^3} :: 1 : 3$; and therefore $m' = \frac{3\,m\,a'^3}{a^3}$. m' is thus found $= \frac{1}{68.5}$, the mass of the Earth being 1.

342. The two tides immediately following one another, or the tides of the day and of the night, should be very unequal when the Sun and Moon are distant from the equator, if the theory of the spheroid, § 336, were just. They are, however, nearly equal; and this has been shewn, by LA PLACE, to be what must necessarily happen in the oscillations of a sea of uniform depth.

The depth of the Sea is therefore nearly the same throughout, or, though not exactly the same, there is a certain mean depth, from which the deviations are not considerable, if we take in a large extent of ocean. If this were not the case, the consecutive tides would not approach so near to equality as they actually do.

343. Great

343. Great extent is necessary, in order that the Sea should be sensibly affected by the action of the Sun and Moon; for it is only by the inequality of that action, on different parts of the mass of waters, that their equilibrium is disturbed; and such inequality cannot take place, unless a great extent of surface be included.

The value of $\cos^2 z$, in the preceding formulas, belonging to different parts of the same sea, must be considerably different, in order that an oscillation of the waters may be produced. The same is true of the horizontal force, of which $\sin 2z$ is the multiplier. This last is at its *maximum* at the distance of 45° from the point where the attracting body is in the zenith.

344. The tides which are experienced in narrow seas, and on shores far from the main body of the ocean, are not produced in those seas by the direct action of the luminaries, but are waves propagated from the great diurnal undulation, and moving with much less velocity.

Of this, the tides in the German Sea, and on the coasts of Britain, are remarkable examples.

The high-water transmitted from the tide in the Atlantic, reaches Ushant between three and four hours after the Moon has passed the meridian, and its ridge stretches NW, so as to fall a little south of the coast of Ireland. This wave soon after divides itself into three; one part passing up the British Channel, another ranging along the west side of Ireland and Scotland, and the third entering the Irish Channel. The first of these flows through the Channel at the rate of about 50 miles an hour, so as to pass through the Straits of Dover, and to reach the Nore about twelve at night. The second, being in a more open sea, moves with more rapidity; by six it has reached the north extremity of the Irish coast; about nine it has got to the Orkney Islands, and forms a ridge or wave extending due north; at twelve, the summit of the same wave extends from the coast of Buchan eastward to the Naze of Norway; and in twelve hours more, it reaches the Nore, where it meets the morning tide, that left the mouth of the Channel only eight hours before. Thus these two tides travel round Britain in about twenty-eight hours, in which time the primitive tide has gone round the whole circumference of the Earth, and nearly 45 degrees more. YOUNG's *Lectures on Natural Philosophy*, vol. I. plate xxxviii. fig. 521.

Thus it is found, that the laws of motion, and the general properties of matter, are the same in the Heavens and on the Earth; that the elliptical motions of the planets, both primary and secondary; the small deviations from those motions, whether in the places of the planets, or in the form and position of their orbits; the facts which concern their figures, their rotation, and the position of their axis; and, lastly, the oscillation of the waters which surround the Earth, are all explained by *one principle*,—That of the mutual Gravitation of all Bodies, with forces directly as their quantities of matter, and inversely as the squares of their distances. The existence of this force, it should be observed, was not assumed as an hypothesis, but was deduced as a necessary consequence from the general facts or laws discovered by KEPLER.

We have thus arrived at the knowledge of a principle which pervades all Nature, and connects together the most distant regions of space, as well as the most remote periods of duration. It is a principle also of singular simplicity. Spherical bodies, made up of particles, attracting according to this law, attract one another according to the same, which could not happen, if the attraction of the particles decreased

in

in any other than the inverse ratio of the squares of the distances. The expression for the force which this law affords, viz. the quantity of matter, divided by the square of the distance, is a line, or a magnitude of one dimension, as the expression of a force must always be when reduced to the utmost simplicity. Lastly, The lines described by two bodies attracting one another according to this law, are always of the second order. No other law could give the same simplicity to the celestial motions, nor is it likely that any one could produce the same stability.

Gravitation, nevertheless, is not conceived by us as a property essential to matter; there may be many other laws equally possible, and the above considerations point out the law actually existing, as one that has been *wisely selected* out of an infinite number.

Is there, then, any physical cause, yet more general, into which Gravitation may be resolved; or is it an ultimate fact, beyond which our knowledge cannot extend? If we look at the ill success of the attempts hitherto made to explain gravitation, we shall be disposed to embrace the latter opinion, and to apply the maxim of BACON: "*Est autem æque imperiti et leviter philosophantis in maxime universalibus causam requirere, ac in subordinatis et subalternis causam non desiderare.*" Nov. Org. lib. I. cap. 48.

If, on the other hand, we consider how many different laws seem to regulate the other phenomena of the material

material world, as in the action of Impulse, Cohesion, Elasticity, Chemical Affinity, Crystallization, Heat, Light, Magnetism, Electricity, Galvanism, the existence of a principle more general than any of these, and connecting all of them with that of Gravitation, appears highly probable.

The discovery of this great principle may be an honour reserved for a future age, and Science may again have to record names which are to stand on the same levels with those of NEWTON and LA PLACE. About such ultimate attainments, it were unwise to be sanguine, and unphilosophical to despair.

END OF VOLUME SECOND.

DIRECTIONS TO THE BINDER.

Vol. II. has four plates to be placed at the end, in the order of the figures.

Plate I. contains from fig. 1. to fig. 7.
II. Fig. 8. to fig. 14.
III. Fig. 15. to fig. 22.
IV. Fig. 23. to fig. 28.

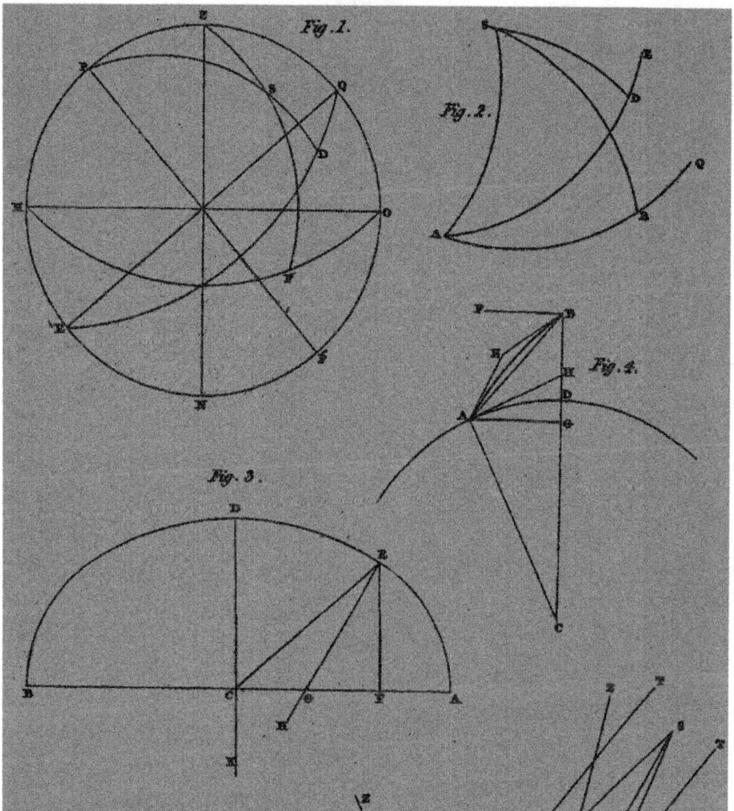
Fig. 1.
Fig. 2.
Fig. 3.
Fig. 4.
Fig. 5.
Fig. 6.
Fig. 7.
Lizars Sculpt Edin.

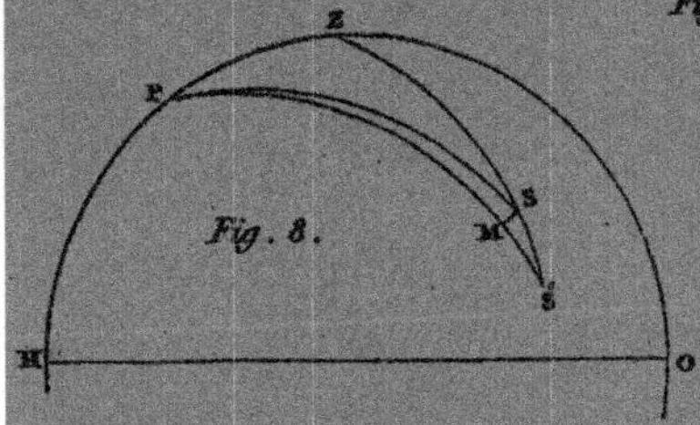

Fig. 8.

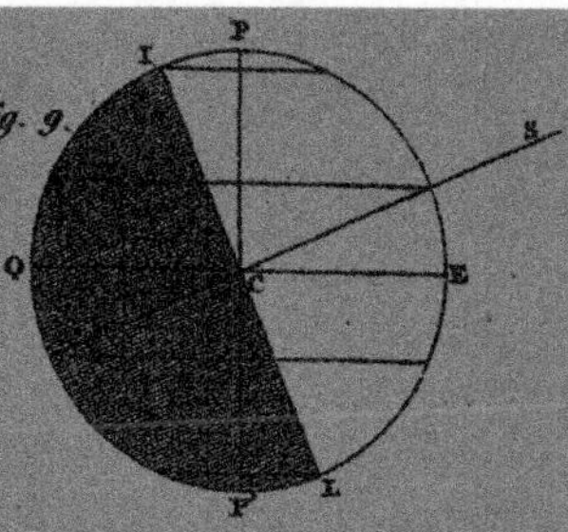

Fig. 9.

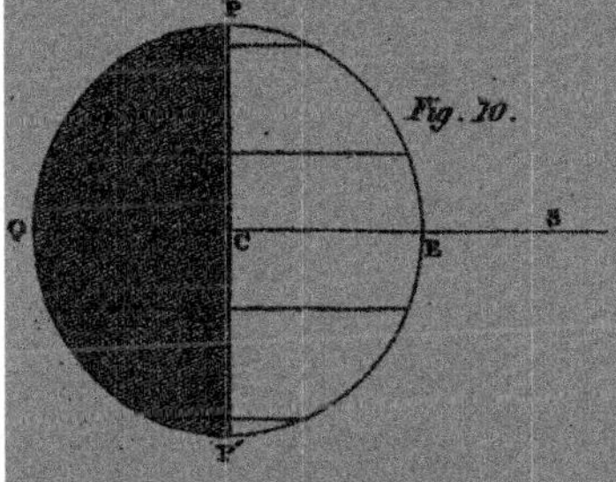

Fig. 10.

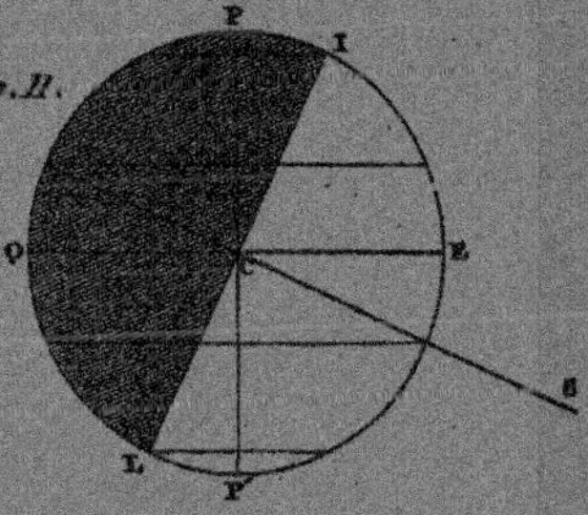

Fig. 11.

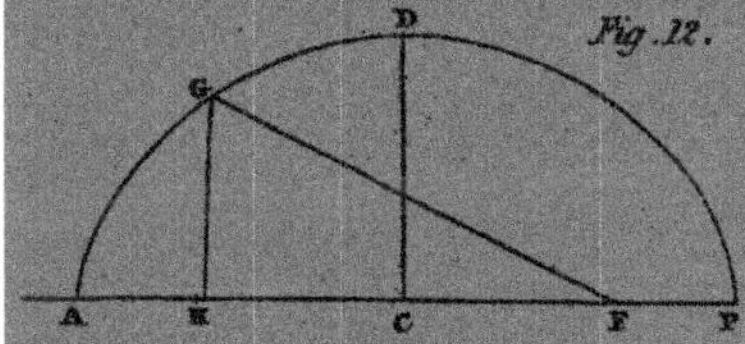

Fig. 12.

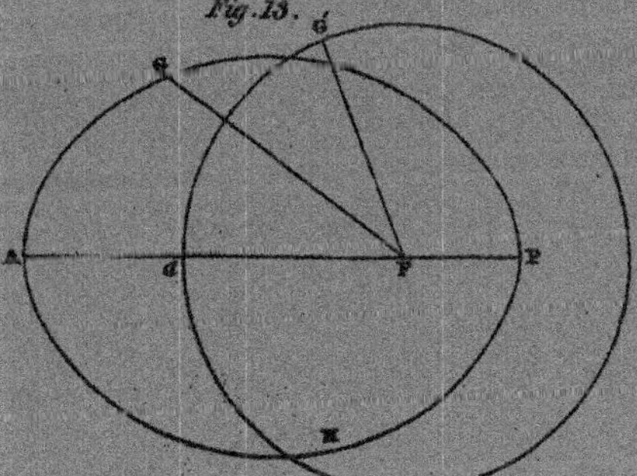

Fig. 13.

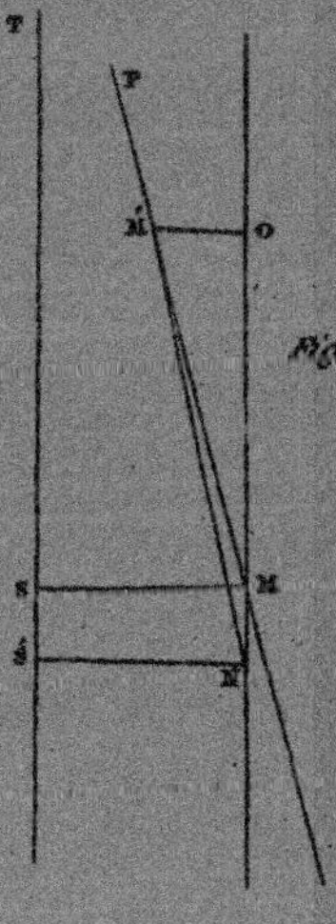

Fig. 14.

Lizars Sculpt. Edin.

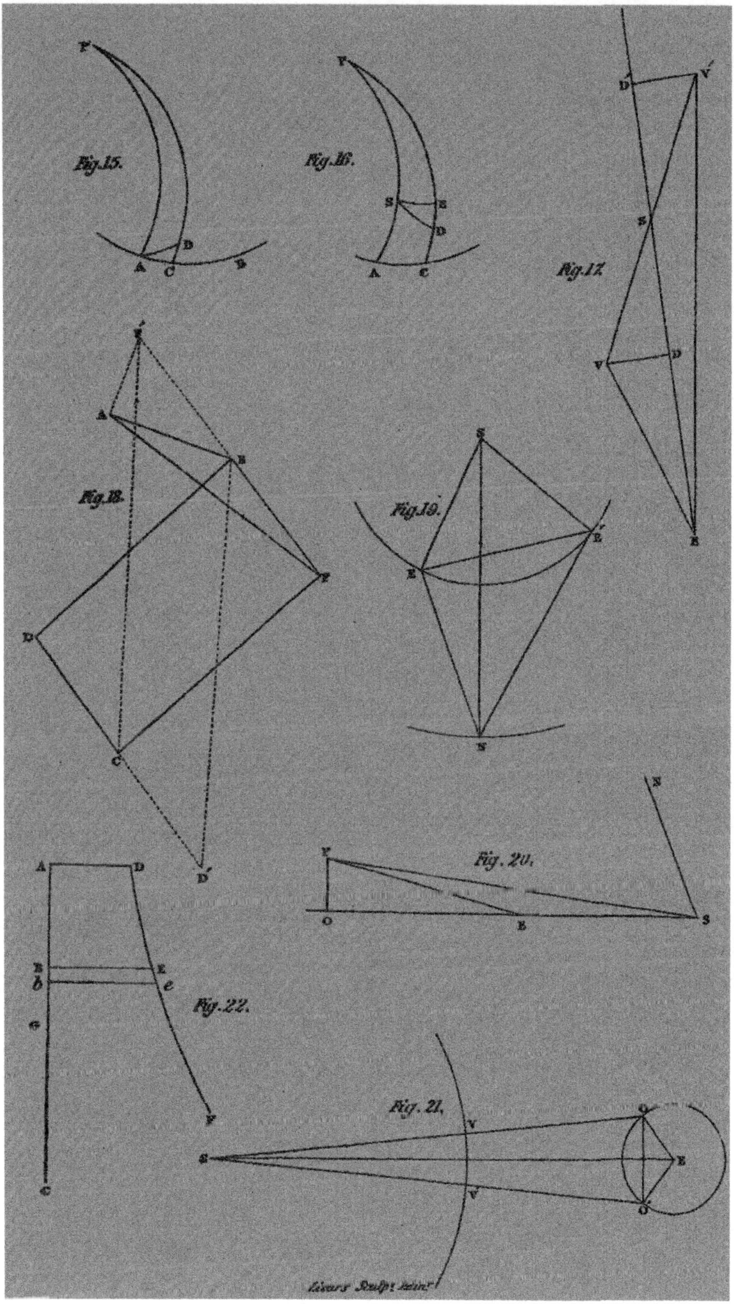
Fig. 15.
Fig. 16.
Fig. 17.
Fig. 18.
Fig. 19.
Fig. 20.
Fig. 21.
Fig. 22.
Lizars Sculpt. Edinr.

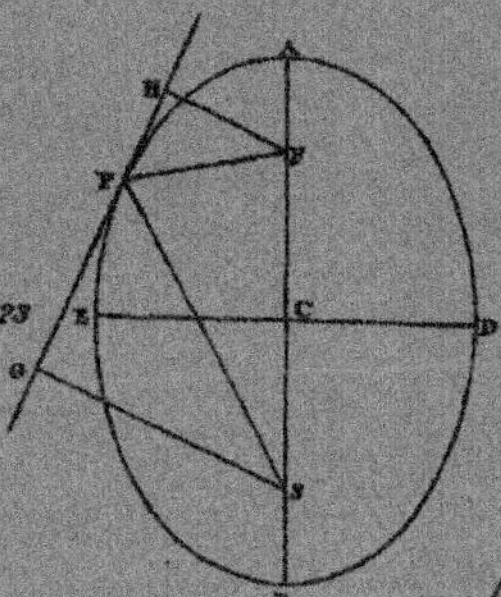

Fig. 23

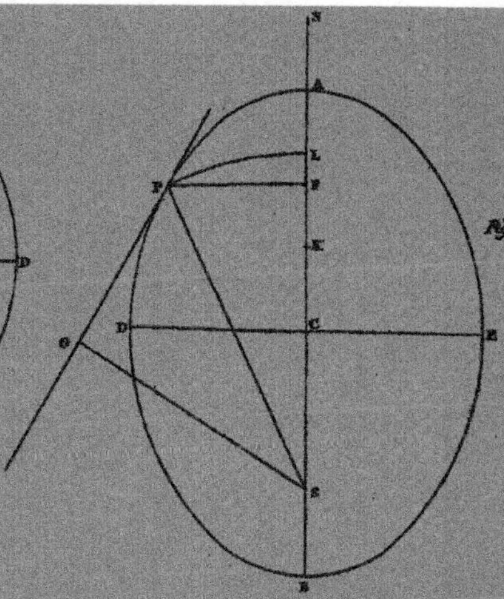

Fig. 24

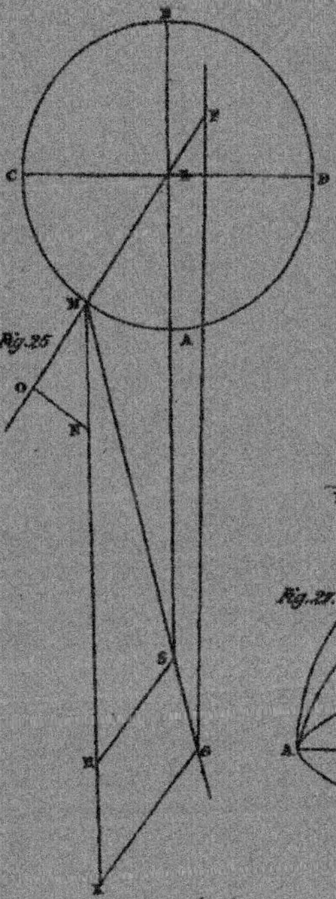

Fig. 25

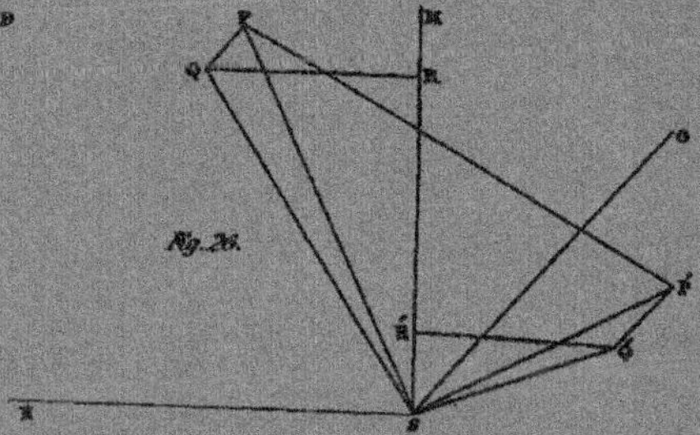

Fig. 26

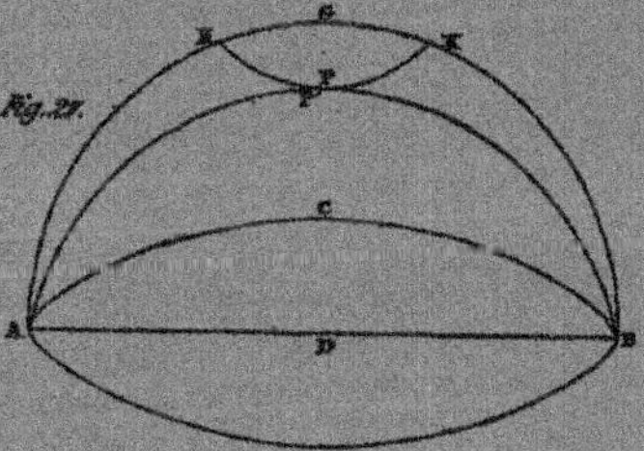

Fig. 27

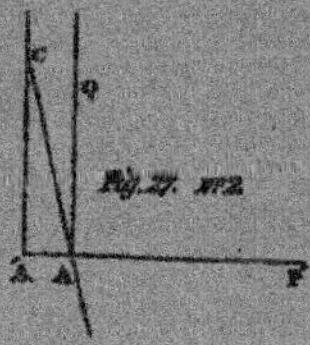

Fig. 27. No. 2.

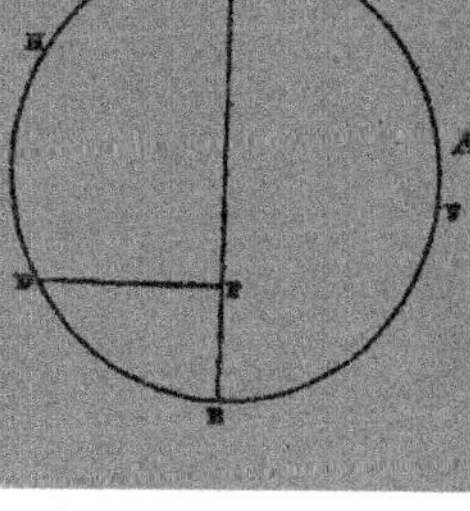

Fig. 28.

Printed in Dunstable, United Kingdom

91219943R00208